Springer Theses

Recognizing Outstanding Ph.D. Research

For further volumes:
http://www.springer.com/series/8790

Aims and Scope

The series "Springer Theses" brings together a selection of the very best Ph.D. theses from around the world and across the physical sciences. Nominated and endorsed by two recognized specialists, each published volume has been selected for its scientific excellence and the high impact of its contents for the pertinent field of research. For greater accessibility to non-specialists, the published versions include an extended introduction, as well as a foreword by the student's supervisor explaining the special relevance of the work for the field. As a whole, the series will provide a valuable resource both for newcomers to the research fields described, and for other scientists seeking detailed background information on special questions. Finally, it provides an accredited documentation of the valuable contributions made by today's younger generation of scientists.

Theses are accepted into the series by invited nomination only and must fulfill all of the following criteria

- They must be written in good English.
- The topic should fall within the confines of Chemistry, Physics, Earth Sciences, Engineering and related interdisciplinary fields such as Materials, Nanoscience, Chemical Engineering, Complex Systems and Biophysics.
- The work reported in the thesis must represent a significant scientific advance.
- If the thesis includes previously published material, permission to reproduce this must be gained from the respective copyright holder.
- They must have been examined and passed during the 12 months prior to nomination.
- Each thesis should include a foreword by the supervisor outlining the significance of its content.
- The theses should have a clearly defined structure including an introduction accessible to scientists not expert in that particular field.

Carsten Brée

Nonlinear Optics in the Filamentation Regime

Doctoral Thesis accepted by
the Humboldt University Berlin, Germany

 Springer

Author
Dr. Carsten Brée
Weierstrass Institute for Applied Analysis
  and Stochastics Leibniz Institute
  in Forschungsverbund Berlin e. V.
Humboldt University
Berlin
Germany

Supervisors
Prof. Uwe Bandelow
Weierstrass Institute for Applied Analysis
  and Stochastics Leibniz Institute
  in Forschungsverbund Berlin e. V.
Humboldt University
Berlin
Germany

Dr. Günter Steinmeyer
Max Born Institute for Nonlinear Optics
  and Short Pulse Spectroscopy
  in Forschungsverbund Berlin e. V.
Berlin
Germany

ISSN 2190-5053   ISSN 2190-5061   (electronic)
ISBN 978-3-642-44358-9   ISBN 978-3-642-30930-4 (eBook)
DOI 10.1007/978-3-642-30930-4
Springer Heidelberg New York Dordrecht London

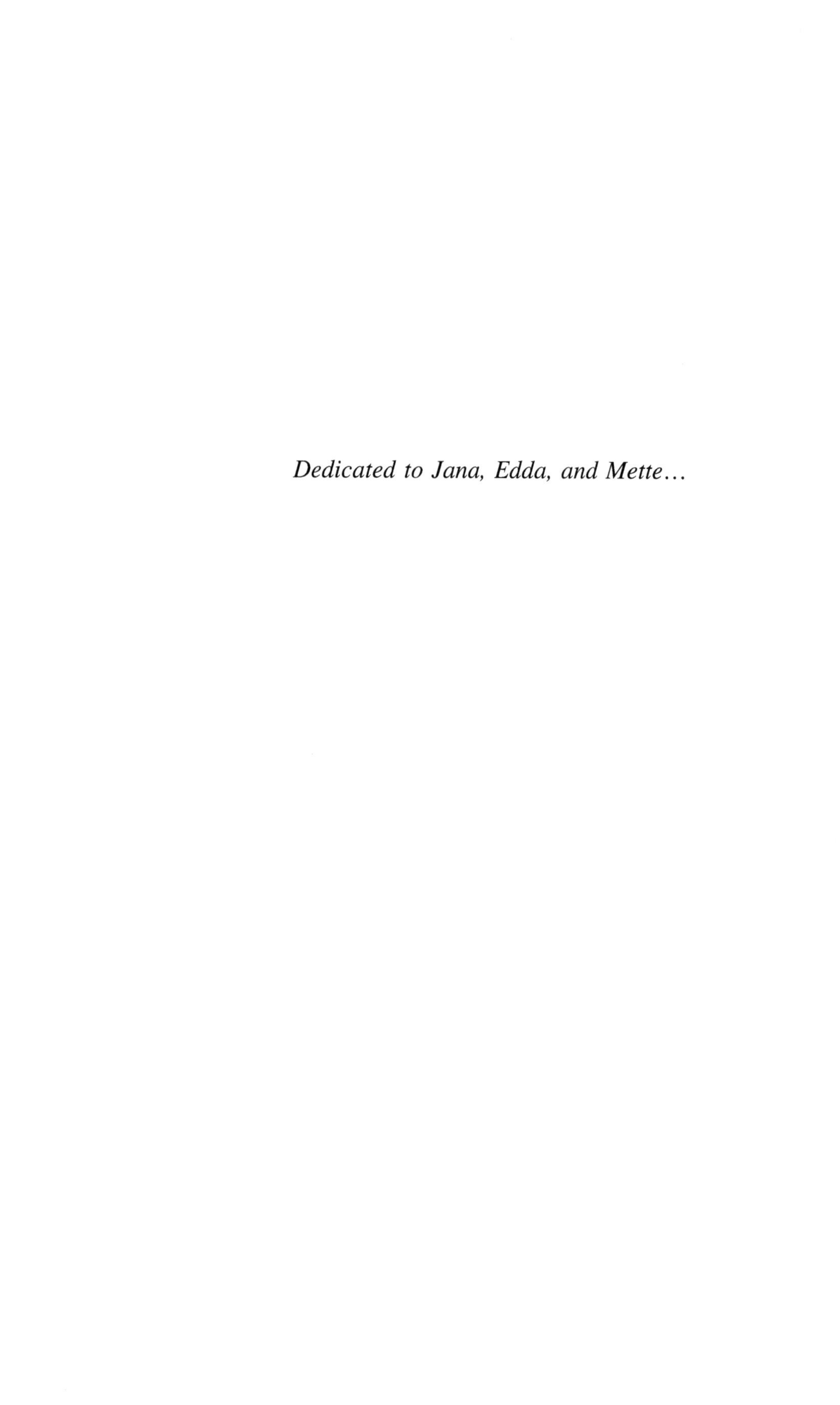

Dedicated to Jana, Edda, and Mette...

Supervisors' Foreword

Ultrashort optical pulses have become a ubiquitous tool in fundamental research, giving access to unprecedented field strength and temporal resolution in the fields of physics and chemistry. Several Nobel prizes in the last decade are immediate witness of this development, most prominently the 2005 Nobel Prize awarded to Hall and Hänsch for the frequency comb.

The concentration of energy on the shortest possible timescale requires an intricate interplay of at least one linear and one nonlinear optical effect. Given the additional role of plasma effects, high-field compression scenarios are significantly more difficult to understand than others. In particular, recent experimental reports on self-compression of optical pulses during propagation in a filament were highly puzzling, challenging theoretical modeling of such compression mechanisms in an unprecedented manner. Filamentary propagation involves spatial as well as temporal effects, both from linear and two competing nonlinear optical effects. This puzzling scenario served as the starting point for the thesis of Carsten Brée. First, with the tangible picture of self-pinching of the intensity distribution of the light, his work delivers a novel explanation for the remarkable and previously unexplained phenomenon of pulse self-compression in filaments. Moreover, the work addresses the scenario of such an intense light pulse experiencing a non-adiabatic change of nonlinearity and dispersion at the interface between a gaseous dielectric material and a solid one.

Finally, and probably most importantly, the thesis delivers a simple and highly practical theoretical approach to estimate the influence of higher order nonlinear optical effects in optics. The results from this thesis shed new light on recent experimental observations, which are controversially discussed because they may completely change our understanding of filamentation, causing a paradigm change concerning the role of higher order nonlinearities in optics. In fact, these results indicate the possibility of plasmaless filaments, enabling dissipationless delivery of highly energetic light pulses in a self-generated guiding geometry.

The thesis has already inspired further work on the nonlinear optical response in various dielectric media, including solids. Other current effort on noble gases target an improved precision of the estimates of saturation and inversion intensity.

It therefore appears that this thesis has already triggered a wave of novel research in this field, trying to confirm (or disprove) the often underestimated role of higher order nonlinearities in optics. As the supervisors, we are therefore extremely pleased that the groundbreaking work of Carsten Brée receives such favorable acknowledgment by publication in the Springer Thesis series.

April 2012   Uwe Bandelow and Günter Steinmeyer

Acknowledgments

I would like to thank Prof. Dr. T. Elsässer and Priv. Doz. Dr. Uwe Bandelow for giving me the opportunity to work on an exciting topic at the interface of experiment and theory.

I am deeply grateful to Prof. Dr. G. Steinmeyer and Dr. A. Demircan for their continuous and valuable scientific advise and for providing respectful and motivating working conditions.

I would also like to thank Prof. Dr. S. Skupin and Dr. L. Bergé for kindly providing the source code for the numerical simulations and for their beneficial contributions to the journal articles published in the course of this Ph.D. project.

Furthermore, I am indebted to Dr. E. T. J. Nibbering who granted access to the laser system used for the experimental parts of this project.

Dipl. Phys. J. Bethge introduced me into the subtleties of few-cycle pulse characterization via spectral phase interferometry, for which I would like to express my honest gratitude.

Dr. Shalva Amiranashvili willingly shared his expertise on envelope descriptions of nonlinear optics, for which I wish to express my deepest gratitude. I also express my gratitude to Dipl. Phys. A. Wilms for discussions about physics beyond one's own nose and for proofreading the manuscript.

My thanks also goes to Dr. C. Grebing, Dipl. Phys. S. Koke, and Dipl. Phys. A. Schmidt for sharing with me their experimental expertise on nonlinear optics.

Finally, I would like to thank all members of the research group "Laser Dynamics" at WIAS and those of the division C2 at MBI for the enjoyable and cooperative working atmosphere.

Contents

List of Publications

Refered publications

1. C. Brée, A. Demircan, and G. Steinmeyer, Asymptotic pulse shapes in filamentary propagation of intense femtosecond pulses , Laser Phys., **19**, 330 (2009).

2. C. Brée, A. Demircan, S. Skupin, L. Bergé, and G. Steinmeyer, Self-pinching of pulsed laser beams during filamentary propagation, Opt. Express, **17**, 16429 (2009).

3. C. Brée, A. Demircan, S. Skupin, L. Bergé, and G. Steinmeyer, Plasma induced pulse breaking in filamentary self-compression, Laser Phys., **20**, 1107 (2010).

4. C. Brée, J. Bethge, S. Skupin, L. Bergé, A. Demircan, G. Steinmeyer, Cascaded self-compression of femtosecond pulses in filaments, New J. Phys. **12**, 093046 (2010).

5. C. Brée, A. Demircan, and G. Steinmeyer, Method for computing the nonlinear refractive index via Keldysh theory, IEEE J. Quantum Electron., **46**, 433 (2010).

6. C. Brée, A. Demircan, and G. Steinmeyer, Modulation instability in filamentary selfcompression, accepted for publication in Laser Phys. (2011)

7. C. Brée, A. Demircan, J. Bethge, E. T. J. Nibbering, S. Skupin, L. Bergé, and G. Steinmeyer, Filamentary pulse self-compression: The impact of the cell windows, Phys. Rev. A **83**, 043803 (2011).

8. J. Bethge, C. Brée, H. Redlin, G. Stibenz, P. Straudt, G. Steinmeyer, A. Demircan, and S. Düsterer, Self-compression of 120 fs pulses in a white-light filament, J. Opt. **13**, 055203 (2011).

9. C. Brée, A. Demircan, and G. Steinmeyer, Saturation of the all-optical Kerr effect, Phys. Rev. Lett. **106**, 183902 (2011).

10. C. Brée, A. Demircan, and G. Steinmeyer, Kramers-Kronig relations and high-order nonlinear susceptibilities, Phys. Rev. A **85**, 033806 (2012).

11. B. Borchers, C. Brée, S. Birkholz, A. Demircan, and G. Steinmeyer, Saturation of the all-optical Kerr effect in solids. Optics Letters **37**, 1541–1543 (2012)

Conference Proceedings

12. C. Krüger, A. Demircan, S. Skupin, G. Stibenz, N. Zhavoronkov, and G. Steinmeyer, Asymptotic pulse shapes in filamentary propagation of femtosecond pulses and self-compression, in: Quantum Electronics and Laser Science Conference, OSA Technical Digest, JTuA44 (2008).

13. C. Krüger, A. Demircan, S. Skupin, G. Stibenz, N. Zhavoronkov, and G. Steinmeyer, Asymptotic pulse shapes and pulse self-compression in femtosecond filaments, in: Ultrafast Phenomena XVI, Proceedings of the 16th International Conference, June 9-13, 2008, Stresa, Italy, Corkum, P., Silvestri, S., Nelson, K.A., Riedle, E., Schoenlein, R.W. (Eds.), Vol. 92 of Springer Series in Chemical Physics, 804 (2009).

14. C. Brée, A. Demircan, S. Skupin, L. Bergé, and G. Steinmeyer, Nonlinear photon z-pinching in filamentary self-compression, in: Conference on Lasers and Electro- Optics/International Quantum Electronics Conference, OSA Technical Digest, ITuC1/1- ITuC1/2 (2009).

15. C. Brée, A. Demircan, S. Skupin, L. Bergé, and G. Steinmeyer, Nonlinear photon z-pinching in filamentary self-compression, in: CLEO/Europe and EQEC 2009 Conference Digest, CF5_4 (2009).

16. C. Brée, J. Bethge, S. Skupin, L. Bergé, A. Demircan, G. Steinmeyer, Double Selfcompression of Femtosecond Pulses in Filaments, in: Quantum Electronics and Laser Science Conference, OSA Technical Digest, JThD6 (2010)

17. C. Brée, J. Bethge, A. Demircan, E. T. J. Nibbering, and G. Steinmeyer, On the Origin of Negative Dispersion Contributions in Filamentary Propagation, in: Conference on Lasers and Electro-Optics, OSA Technical Digest, CMU2 (2010).

18. C. Brée, G. Steinmeyer, and A. Demircan, Saturation of the all-optical Kerr effect, in CLEO:2011 - Laser Applications to Photonic Applications, OSA Technical Digest (CD) (Optical Society of America, 2011), paper CWR6.

19. C. Brée, A. Demircan, and G. Steinmeyer, Saturation of the All-Optical Kerr Effect, in CLEO/Europe and EQEC 2011 Conference Digest, OSA Technical Digest (CD) (Optical Society of America, 2011), paper EF4 6.

Symbols and Conventions

$\mathrm{Re}z$	Real part of z
$\mathrm{Im}z$	Imaginary part of z
$\partial_x, \frac{\partial}{\partial x}$	Partial differentiation with respect to x
$\vec{\nabla}$	Nabla operator $\vec{\nabla} = \partial_x \vec{e}_x + \partial_y \vec{e}_y + \partial_z \vec{e}_y$
$\vec{e}_x$	Unit vector in x direction
$\vec{r}$	Position vector $\vec{r} = x\vec{e}_x + y\vec{e}_y + z\vec{e}_z$
Δ	Laplace operator $\Delta = \partial_x^2 + \partial_y^2 + \partial_z^2$
$\Delta_\perp$	Transverse Laplacian $\Delta_\perp = \partial_x^2 + \partial_y^2$
$\delta(x)$	Delta distribution $\int dx\, \delta(x) f(x) = f(0)$
$\Theta(x)$	Heaviside function $\Theta(x) = \begin{cases} 1, & \text{if } x > 0 \\ 0, & \text{if } x < 0 \end{cases}$
$\mathrm{sgn}(x)$	sign function $\mathrm{sgn}(x) = \begin{cases} 1, & \text{if } x > 0 \\ 0, & \text{if } x = 0 \\ -1, & \text{if } x < 0 \end{cases}$
$\widehat{G}(\omega) = \mathcal{F}[G](\omega)$	Fourier transform, $\widehat{G}(\omega) = \frac{1}{2\pi} \int G(t) e^{i\omega t} dt$
$G(t) = \mathcal{F}^{-1}\left[\widehat{G}\right](t)$	Inverse Fourier transform, $G(t) = \int \widehat{G}(\omega) e^{-\omega t} d\omega$
$\mathcal{H}[f](\omega)$	Hilbert transform, $\mathcal{H}[f](\omega) = -\frac{1}{\pi} P \int\limits_{-\infty}^{\infty} \frac{f(\Omega)}{\Omega - \omega} d\Omega$
ϵ_0	Vacuum permittivity $\epsilon_0 = 8.854187817 \times 10^{-12}$ As/Vm
μ_0	Vacuum permeability $\mu_0 = 4\pi \times 10^{-7}$ Vs/Am
ϵ	Relative permittivity
ω	Angular frequency
ω_0	Center angular frequency
c	Velocity of light in vacuum, $c = 299792458$ m/s
$\vec{k}$	Wavevector $\vec{k} = k_x \vec{e}_x + k_y \vec{e}_y + k_z \vec{e}_z$
$k(\omega)$	Norm of wavevector, $k(\omega) = \sqrt{k_x^2 + k_y^2 + k_z^2} = n(\omega)\omega/c$

k_0	Wave number at center frequency ω_0, $k_0 = k(\omega_0)$
$n(\omega)$	Refractive index at frequency ω
n_0	Refractive index $n(\omega_0)$ at center frequency ω_0
$\alpha(\omega)$	Absorption coefficient at frequency ω
α_0	Absorption coefficient $\alpha(\omega_0)$ at center frequency ω_0
n_{2k}	$2k$-th order Kerr coefficient
σ_K	Cross section for K-photon ionization
β_K	Nonlinear coefficient for K-photon absorption
ω_0	Beam waist
t_p	Pulse duration
P_cr	Critical power for self-focusing, $P_\mathrm{cr} \approx \lambda^2/2\pi n_0 n_2$
$n(I)$	Intensity dependent refractive index, $n(I) = \Sigma_{k \geq 0} n_{2k} I^k$
$\Delta n(I)$	Nonlinearly induced refractive index change, $\Delta n(I) = n(I) - n_0$
$\Delta\alpha(I)$	Nonlinearly induced change of absorption, $\Delta\alpha(I) = \alpha(I) - \alpha_0$
β_n	n-th order dispersion coefficient at center frequency. ω_0, $\beta_n = \partial_\omega^n k(\omega)\vert_{\omega_0}$
ρ_0	neutral density
IDRI	Intensity dependent refractive index
FWHM	Full Width Half Maximum
NLSE	Nonlinear Schrödinger Equation
NEE	Nonlinear Envelope Equation
SVEA	Slowly Varying Envelope Approximation
SEWA	Slowly Evolving Wave Approximation
GVD	Group Velocity Dispersion
TOD	Third Order Dispersion
MPI	Multiphoton Ionization
MPA	Multiphoton Absorption
HT	Hilbert Transform
FFT	Fast Fourier Transform
TDSE	Time Dependent Schrödinger Equation

Chapter 1
Introduction

Femtosecond filaments are narrow self-confined beams of laser light maintaining their beam diameters over distances widely exceeding the classical Rayleigh range of a laser beam [1]. As shown in Fig. 1.1, such a self-organized, filamentary structure of light and free electric charges emerges when pulsed femtosecond laser radiation is focused into a gas cell. In the setup depicted, the cell was filled with argon at atmospheric pressure. Due to the high optical intensities involved, the gas is ionized to form a dilute plasma, as is evident by the characteristic bluish-violet fluorescence. The latter sets in at the position of the so-called nonlinear focus F_{NL}, which actually precedes the position of the linear focus F. By the existence both of a nonlinear focal point and the aforementioned long range propagation property, filaments apparently defy the diffraction laws of linear optics [2, 3]. Indeed, the physical mechanisms behind filament formation have given rise to many controversial discussions on their sub-diffractive nature. Quite clearly, a filament is a highly nonlinear optical phenomenon that can only be generated with pulsed laser beams at peak powers in the GW and MW range in gases and condensed media, respectively [4, 5]. In liquids, filaments were first observed by Pilipetskii and Rustamov [6]. In gases, which possess a Kerr nonlinearity three orders of magnitude smaller than that of liquids, the first observation of atmospheric laser filaments [1] became only possible three decades later with the development of the chirped pulse amplification (CPA) technique [7] in the mid-eighties, which provides ultrashort laser pulses up to the PW level. With typically observed filament diameters of some $100\,\mu$m in gases, the peak power of some 10 GW required to observe filamentation translates into field strengths of several 10^{10} V/m. This value approaches inner-atomic binding forces of the order of one atomic unit of the electric field strength (5.14×10^{11} V/m).

Possible applications of femtosecond filamentation are widespread. Spectral broadening due to self-phase modulation in filaments was first observed in Ref. [8]. The resulting white-light supercontinua have found applications in optical coherence tomography or white-light LIDAR (light detection and ranging). The latter involves atmospheric analysis utilizing the white-light continuum within a filament generated by a terawatt laser source, as realized, e.g., within the Teramobile project [9].

C. Brée, *Nonlinear Optics in the Filamentation Regime*, Springer Theses, 1
DOI: 10.1007/978-3-642-30930-4_1, © Springer-Verlag Berlin Heidelberg 2012

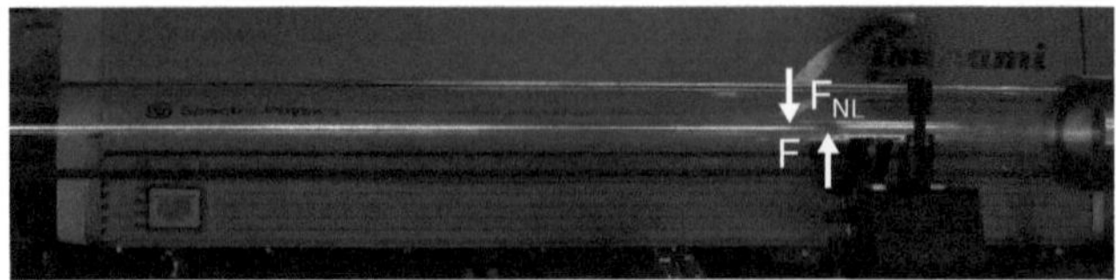

Fig. 1.1 A femtosecond filament obtained by loosely focusing pulsed femtosecond laser radiation into an argon filled gas cell. Initial pulse energy and duration are about 1 mJ and 45 fs, respectively. F and F_{NL} denote linear and nonlinear focus, respectively

A further example of remote optical sensing is provided by laser induced breakdown spectroscopy (LIBS) in conjunction with the long-range propagation of femtosecond filaments [10]. With this technique, remote samples can be irradiated with intensities sufficiently strong to trigger photoionization, which enables the detection of characteristic atomic emission lines. This method can, e.g., be used for a remote analysis of objects of cultural heritage, e.g., sculptures or monuments [11]. Further potential applications of femtosecond filamentation involve the wireless transfer of electric current or the generation of Terahertz radiation [12]. The former aims, e.g., at a contactless pantograph for the power supply of high-speed trains [13], while the latter can be exploited, e.g., for security screening issues such as the remote sensing of explosives.

The intensities within filaments are sufficiently high to create a dilute plasma via multiphoton or tunneling ionization processes. This leads to a competition between cumulative effects from plasma contributions to the refractive index and instantaneous Kerr contributions, which yields a considerable dynamics of the optical pulse shape both, in the spatial and temporal domain [14]. Under suitable experimental conditions, this may lead to the surprising experimental effect of pulse self-compression: in addition to the spatial confinement within the laser beam, nonlinear optical effects have been shown to produce further pulse shortening upon nonlinear propagation, coming close to the old dream of the formation of optical light bullets [15], which ideally self-concentrate their energy while traveling through a nonlinear medium. This phenomenon, while analyzed previously in numerical simulations [16], is now shown to be chiefly of spatial nature, strongly contrasting any previous method for pulse compression [17, 18]. Furthermore, it is shown both experimentally and in numerical simulations that the processes leading to pulse self-compression can be cascaded for suitably chosen input pulse parameters [19].

Filaments are strong nonlinear attractors and have been shown to possess self-healing capabilities of their transverse spatial profile. The time domain analogue of this self-restoration property is another surprising finding with pulse self-compression. It is related to the fact that experimental set-ups frequently require cell windows to employ the nonlinearity of atomic gases. Compared to the dispersion and nonlinearity of a gaseous medium, cell windows represent a sudden and non-adiabatic change in either property of a factor of some thousand. Either the nonlinear or the dispersive linear influence should immediately destroy the short temporal signature of the self-compressed pulses generated within the filament. This puzzling controversy was recently theoretically solved [20, 21], predicting that a self-healing mechanism may regenerate the original shortness of the pulse; a prediction that will now be experimentally verified within this thesis [22].

Finally, with field strengths approximating inner-atomic binding forces, the question arises whether there is an influence of higher-order Kerr effects beyond $\chi^{(3)}$ in any of the above-discussed. As self-focusing in a Kerr medium with $\chi^{(3)} > 0$ leads to an intensity blow-up, counteracting higher-order Kerr terms have originally been discussed phenomenologically to explain filament formation as such [23]. On the contrary, recent papers have mostly neglected nonlinear susceptibilities higher than third order [4, 5]. Instead, clamping of $\chi^{(3)}$-governed nonlinear refraction is explained by Drude contributions from plasma formation. This perspective has recently been challenged by measurements [24, 25] of the nonlinearly induced birefringence, clearly indicating an effect of higher-order Kerr contributions to filament formation, which gave rise to a controversial discussion [26–32]. This thesis provides a totally independent approach to the question of a paradigm change in the explanation of femtosecond filaments, computing the nonlinear refractive index changes via a Kramers-Kronig transformation of multiphoton ionization rates [33]. For the case of two-photon ionization, which is related to second-order nonlinear refraction governed by $\chi^{(3)}$ [34–36], this method yields results which are in excellent agreement with all available accepted experimental and theoretical materials published previously.

As a single-parameter theory, depending only on the ionization energy of the respective atom, the employed model provides estimates on the nonlinear refractive index that clearly confirm the importance of the higher-order Kerr coefficients [24, 25] for filament stabilization [31]. This opens a perspective on a paradigm change in the understanding of nonlinear optics at extreme intensities. Quite clearly, under certain conditions, high-order refractive nonlinearities may all set in simultaneously above a threshold intensity, similar to the behavior of dissipative nonlinearities in phenomena as exploited in high-harmonic generation. This may open a completely new perspective at nonlinear optical phenomena in the extreme, in the highly interesting merger region between traditional perturbative nonlinear optics and high-field nonlinear optics.

The numerical simulations discussed in this thesis were performed with a FORTRAN90 code which was kindly provided by Dr. Luc Bergé (Commissariat à l'Energie Atomique et aux Energies Alternatives, Arpajon, France) and Prof. Stefan Skupin (Max Planck Institute for the Physics of Complex Systems, Dresden, and Friedrich Schiller University, Institute of Condensed Matter Theory and Optics, Jena). The code uses the Message Passing Interface (MPI) libraries which enable parallel computation.

This thesis arose from a joint project of Weierstrass Institute for Applied Analysis and Stochastics (WIAS) and Max Born Institute for Nonlinear Optics and Short Pulse Spectroscopy (MBI). All simulations were performed on the blade cluster `euler` (Hewlett-Packard CP3000BL) at WIAS. For the experimental parts of the thesis, a Ti:sapphire regenerative amplifier system (Spectra Physics Spitfire) at MBI was employed. Access to the laser system was kindly granted by Dr. Erik Nibbering.

References

1. A. Braun, G. Korn, X. Liu, D. Du, J. Squier, G. Mourou, Self-channeling of high-peak-power femtosecond laser pulses in air. Opt. Lett. **20**, 73 (1995)
2. S.L. Chin, Y. Chen, O. Kosareva, V.P. Kandidov, F. Théberge, What is a filament? Laser Phys. **18**, 962 (2008)
3. M. Born, E. Wolf, *Principles of Optics* (Cambridge University Press, Cambridge, 1999)
4. L. Berge, S. Skupin, R. Nuter, J. Kasparian, J.P. Wolf, Ultrashort filaments of light in weakly ionized, optically transparent media. Rep. Prog. Phys. **70**, 1633 (2007)
5. A. Couairon, A. Mysyrowicz, Femtosecond filamentation in transparent media. Phys. Rep. **441**, 47 (2007)
6. N.F. Pilipetskii, A.R. Rustamov, Observation of selffocusing of light in liquids. JETP Lett. **2**, 88 (1965)
7. D. Strickland, G. Mourou, Compression of amplified chirped optical pulses. Opt. Commun. **56**, 219 (1985). ISSN 0030–4018
8. R.R. Alfano, S.L. Shapiro, Observation of self-phase modulation and small-scale filaments in crystals and glasses. Phys. Rev. Lett. **24**, 592 (1970)
9. H. Wille, M. Rodriguez, J. Kasparian, D. Mondelain, J. Yu, A. Mysyrowicz, R. Sauerbrey, J.P. Wolf, L. Wöste, Teramobile: a mobile femtosecond-terawatt laser and detection system. Eur. Phys. J. Appl. Phys. **20**, 183 (2002)
10. K. Stelmaszczyk, P. Rohwetter, G. Mejean, J. Yu, E. Salmon, J. Kasparian, R. Ackermann, J.P. Wolf, L. Wöste, Long-distance remote laser-induced breakdown spectroscopy using filamentation in air. Appl. Phys. Lett. **85**, 3977 (2004)
11. S. Tzortzakis, D. Anglos, D. Gray, Ultraviolet laser filaments for remote laser-induced breakdown spectroscopy (LIBS) analysis: applications in cultural heritage monitoring. Opt. Lett. **31**, 1139 (2006)
12. I. Babushkin, W. Kuehn, C. Köhler, S. Skupin, L. Bergé, K. Reimann, M. Woerner, J. Herrmann, T. Elsässer, Ultrafast spatiotemporal dynamics of terahertz generation by ionizing two-color femtosecond pulses in gases. Phys. Rev. Lett. **105**, 053903 (2010)
13. A. Houard, C. D'Amico, Y. Liu, Y.B. Andre, M. Franco, B. Prade, A. Mysyrowicz, E. Salmon, P. Pierlot, L.-M. Cleon, High current permanent discharges in air induced by femtosecond laser filamentation. Appl. Phys. Lett. **90**, 171501 (2007)
14. M. Mlejnek, E.M. Wright, J.V. Moloney, Dynamic spatial replenishment of femtosecond pulses propagating in air. Opt. Lett. **23**, 382 (1998)
15. Y. Silberberg, Collapse of optical pulses. Opt. Lett. **15**, 1282 (1990)
16. S. Skupin, G. Stibenz, L. Berge, F. Lederer, T. Sokollik, M. Schnürer, N. Zhavoronkov, G. Steinmeyer, Self-compression by femtosecond pulse filamentation: experiments versus numerical simulations. Phys. Rev. E **74**, 056604 (2006)
17. C. Brée, A. Demircan, S. Skupin, L. Bergé, G. Steinmeyer, Self-pinching of pulsed laser beams during filamentary propagation. Opt. Express **17**, 16429 (2009)
18. C. Brée, A. Demircan, S. Skupin, L. Bergé, G. Steinmeyer, Plasma induced pulse breaking in filamentary self compression. Laser Phys. **20**, 1107 (2010)
19. C. Brée, J. Bethge, S. Skupin, L. Bergé, A. Demircan, G. Steinmeyer, Cascaded self-compression of femtosecond pulses in filaments. New J. Phys. **12**, 093046 (2010)
20. L. Berge, S. Skupin, G. Steinmeyer, Temporal self-restoration of compressed optical filaments. Phys. Rev. Lett. **101**, 213901 (2008)
21. L. Bergé, S. Skupin, G. Steinmeyer, Self-recompression of laser filaments exiting a gas cell. Phys. Rev. A **79**, 033838 (2009). doi:10.1103/PhysRevA.83.043803

22. C. Brée, A. Demircan, J. Bethge, E.T.J. Nibbering, S. Skupin, L. Bergé, G. Steinmeyer, Filamentary pulse self-compression: the impact of the cell windows. Phys. Rev. A **83**, 043803 (2011). doi:10.1103/PhysRevA.83.043803
23. I.G. Koprinkov, A. Suda, P. Wang, K. Midorikawa, Self-compression of high-intensity femtosecond optical pulses and spatiotemporal soliton generation. Phys. Rev. Lett. **84**, 3847 (2000)
24. V. Loriot, E. Hertz, O. Faucher, B. Lavorel, Measurement of high order Kerr refractive index of major air components. Opt. Express **17**, 13429 (2009)
25. V. Loriot, E. Hertz, O. Faucher, B. Lavorel, Measurement of high order Kerr refractive index of major air components: erratum. Opt. Express **18**, 3011 (2010)
26. A. Teleki, E.M. Wright, M. Kolesik, Microscopic model for the higher-order nonlinearity in optical filaments. Phys. Rev. A **82**, 065801 (2010)
27. M. Kolesik, E.M. Wright, J.V. Moloney, Femtosecond filamentation and higher-order nonlinearities. Opt. Lett. **35**, 2550 (2010)
28. M. Kolesik, D. Mirell, J.-C. Diels, J.V. Moloney, On the higher-order Kerr effect in femtosecond filaments. Opt. Lett. **35**, 3685 (2010)
29. Y.H. Chen, S. Varma, T.M. Antonsen, H.M. Milchberg, Direct measurement of the electron density of extended femtosecond laser pulse-induced filaments. Phys. Rev. Lett. **105**, 215005 (2010)
30. W. Ettoumi, P. Béjot, Y. Petit, V. Loriot, E. Hertz, O. Faucher, B. Lavorel, J. Kasparian, J.-P. Wolf, Spectral dependence of purely-Kerr-driven filamentation in air and argon. Phys. Rev. A **82**, 033826 (2010)
31. P. Bejot, J. Kasparian, S. Henin, V. Loriot, T. Vieillard, E. Hertz, O. Faucher, B. Lavorel, J.-P. Wolf, Higher-order Kerr terms allow ionization-free filamentation in gases. Phys. Rev. Lett. **104**, 103903 (2010)
32. J. Kasparian, P. Béjot, J.-P. Wolf, Arbitrary-order nonlinear contribution to self-steepening. Opt. Lett. **35**, 2795 (2010)
33. C. Brée, A. Demircan, G. Steinmeyer, Saturation of the all-optical Kerr effect. Phys. Rev. Lett. **106**, 183902 (2011). doi:10.1103/PhysRevLett.106.183902
34. M. Sheik-Bahae, D.J. Hagan, E.W. van Stryland, Dispersion and band-gap scaling of the electronic Kerr effect in solids associated with two-photon absorption. Phys. Rev. Lett. **65**, 96 (1990)
35. M. Sheik-Bahae, D.C. Hutchings, D.J. Hagan, E.W. van Stryland, Dispersion of bound electronic nonlinear refraction in solids. IEEE J. Quantum Electron. **27**, 1296 (1991)
36. C. Brée, A. Demircan, G. Steinmeyer, Method for computing the nonlinear refractive index via Keldysh theory. IEEE J. Quantum Electron. **4**, 433 (2010)

Chapter 2
Theoretical Foundations of Femtosecond Filamentation

In the following chapter, the theoretical modeling of femtosecond filamentation is discussed. For a detailed understanding of this phenomenon, the dynamical equation governing the evolution of the laser electric field have to be identified. As only femtosecond filaments in gases are considered here, these are provided by Maxwell's equations [1] in an isotropic, homogeneous, non magnetizable dielectric. Thermal effects can be neglected here as they do not show up on a femtosecond timescale. In addition, the propagation equations admit further simplification as the radiation emitted by modern laser sources exhibits a highly directional character. In the following, the positive z-direction is chosen as the propagation direction of the beam. The electric field is decomposed into plane waves with wave vector $\vec{k}$. Then, the notion of directional beam propagation along z implies $k_z > 0$ and $k_\perp/|\vec{k}| \ll 1$, where $k_\perp$ is the modulus of the transverse wave vector, i.e., $k_\perp = \sqrt{k_x^2 + k_y^2}$. With these assumptions, it can be shown that with good accuracy, the Maxwell equations can be factorized [2, 3], yielding a first order partial differential equation in z, also known as Forward Maxwell's Equation (FME) [4]. The latter governs the evolution of the directional laser field. Compared to Maxwell's equations, the FME allows for a greatly simplified numerical treatment and speeds up the calculations. Moreover, the FME allows the description of ultra-short, ultra-broadband laser radiation emitted by modern, mode-locked femtosecond laser sources. The latter emit laser pulses with durations $<10\,\mathrm{fs}$. For laser radiation with a spectrum centered around $800\,\mathrm{nm}$, this duration corresponds to less than three oscillations of the optical carrier wave. While the propagation of narrow-band optical pulses in a Kerr medium can be adequately described by a Nonlinear Schrödinger Equation (NLSE) [5], the slowly varying envelope approximation (SVEA) fails for these few-cycle pulses. Nevertheless, assuming moderate restrictions on the pulse and the propagation medium, a nonlinear envelope equation (NEE) [6] can be derived. The NEE is a generalization of the NLSE of Ref. [5] and turned out to be a successful model describing the dynamics of few-cycle femtosecond pulses, reproducing experimental results [7]. In fact, historically, the NEE may be considered an ancestor of the more general FME.

C. Brée, *Nonlinear Optics in the Filamentation Regime*, Springer Theses, DOI: 10.1007/978-3-642-30930-4_2, © Springer-Verlag Berlin Heidelberg 2012

A complete description of intense laser radiation propagating in a dielectric medium further requires an appropriate modeling of the polarization $\vec{P}$ due to the response of bound electrons induced by the laser field. As in filamentation intensities of the order of $10^{13}\,\text{W/cm}^2$ are involved [8], the polarization is expected to depend on the electric field in a nonlinear manner. Moreover, the intensity levels achieved in filamentation experiments are sufficiently high to ionize the medium, resulting in the generation of a dilute plasma. This gives rise to a non-zero electron density ρ and an electron current density $\vec{J}$ coupling to the electric field. However, the laser wavelength typically used for the generation of femtosecond filaments is 800 nm (the characteristic wavelength of a Ti:sapphire amplifier), corresponding to a photon energy of $\approx 1.55\,\text{eV}$. In contrast, the ionization potential of the gases relevant for filamentation experiments varies between 10 and 25 eV. This suggests that ionization does not proceed via direct (single-photon) photoionization. Rather, ionization proceeds in a highly nonlinear manner, via, e.g., multiphoton or tunneling ionization [9, 10] which leads to an equally nonlinear dependence of ρ and $\vec{J}$ on the laser electric field.

Finally, the aim is to identify those mechanisms leading to the observed long-range propagation [11–13] of femtosecond filaments as well as other characteristic properties as will be detailed below. To this aim, the envelope equation is analyzed in certain limiting cases in order to isolate the dominant effects contributing to the specific phenomenon under consideration.

2.1 The Forward Maxwell Equations

Maxwell's equations governing the evolution of an electromagnetic field in a dielectric material may be expressed as a coupled set of vector-wave equations for the electric field $\vec{E}$, the dielectric displacement $\vec{D}$ and the current density $\vec{J}$ according to [1, 8, 14]

$$\vec{\nabla}(\vec{\nabla} \cdot \vec{E}) - \vec{\nabla}^2 \vec{E} = -\mu_0 \left(\frac{\partial^2 \vec{D}}{\partial t^2} + \frac{\partial \vec{J}}{\partial t} \right) \tag{2.1}$$

$$\vec{\nabla} \cdot \vec{D} = \rho. \tag{2.2}$$

Here, $\vec{D} = \epsilon_0 \vec{E} + \vec{P}$ is the dielectric displacement which accounts for the bound-charge density due to the polarization $\vec{P}$ induced by the laser electric field. The polarization corresponds to an ensemble average of the atomic or molecular dipole moments induced by the laser field. Throughout the thesis, the paraxial approximation is used, assuming that the laser beam may be Fourier decomposed into plane waves with wave-vectors $\vec{k}$ satisfying

$$k_\perp \ll |\vec{k}|, \tag{2.3}$$

such that the angle between $\vec{k}$ and the optical axis is sufficiently small. As discussed in the introductory remarks to this chapter, this is a reasonable assumption as laser beams exhibit a highly directional character and low beam divergences. Moreover, the polarization is decomposed according to

$$\vec{P} = \vec{P}^{(1)} + \vec{P}_{\mathrm{NL}},\tag{2.4}$$

where the first term $\vec{P}^{(1)}$ varies linearly and the second term varies nonlinearly with the electric field. Thus, $\vec{P}^{(1)}$ describes classical, linear optical phenomena, while the nonlinear response $\vec{P}_{\mathrm{NL}}$ leads to nonlinear optical effects and induces self-interactions of the optical field.

For an isotropic, homogeneous medium, $\vec{P}^{(1)}$ is collinear to the electric field. In the following, it is often useful to treat Eq. (2.2) in the frequency-domain representation. The frequency-domain analogue $\widehat{G}(\omega)$ is related to the function $G(t)$ via the Fourier-transform $\mathcal{F}$, for which the following convention is adopted throughout the thesis,

$$\widehat{G}(\omega) = \mathcal{F}[G](\omega) \equiv \frac{1}{2\pi} \int G(t) e^{i\omega t} dt \tag{2.5}$$

$$G(t) = \mathcal{F}^{-1}[\widehat{G}](t) \equiv \int \widehat{G}(\omega) e^{-i\omega t} d\omega. \tag{2.6}$$

Assuming local response,[1] the frequency domain representation of the linear polarization may be written as [14]

$$\widehat{\vec{P}}^{(1)}(\vec{r}, \omega) = \epsilon_0 \chi^{(1)}(\omega) \widehat{\vec{E}}(\vec{r}, \omega). \tag{2.7}$$

The first order susceptibility $\chi^{(1)}$ is related to the frequency dependent refractive index $n(\omega)$ and absorption coefficient $\alpha(\omega)$ via $(n(\omega)+i\alpha(\omega)c/2\omega)^2 = \epsilon(\omega)$, where the dielectric permittivity is given by the relation $\epsilon(\omega) = 1 + \chi^{(1)}(\omega)$. It has been shown in Refs. [3, 8] that the approximation $\vec{\nabla} \cdot \vec{E} \approx 0$ is justified if, in addition to the paraxiality criterion Eq. (2.3), the nonlinear polarization satisfies the inequality

$$\frac{|P_{\mathrm{NL},i}|}{\epsilon_0 n^2(\omega)} \ll |E_i|, \tag{2.8}$$

where $k(\omega) := |\vec{k}| = n(\omega)\omega/c$ describes the modulus of the wave vector and $i = x, y, z$ labels the vector components. Thus, exploiting the condition Eq. (2.8), the frequency domain analogue of Eq. (2.2) reads

$$\frac{\partial^2 \widehat{\vec{E}}}{\partial z^2} + k^2(\omega)\widehat{\vec{E}} + \nabla_\perp^2 \widehat{\vec{E}} = -\mu_0 \omega^2 \left(P_{\mathrm{NL}} + i\frac{\widehat{J}}{\omega} \right), \tag{2.9}$$

---

[1] Nonlocally responding media play a crucial role for the physics of negative refraction [15]. In these media, the susceptibility $\chi^{(1)}(\omega, \vec{k})$ depends both on the frequency ω and the wave vector $\vec{k}$. The non-local analogue to Eq. (2.7) therefore involves a convolution in the spatial domain.

where the imaginary part of the linear susceptibility has been neglected, i.e. $k^2(\omega) = \omega^2 \epsilon(\omega)/c^2$, with a **real-valued** dielectric function $\epsilon(\omega)$. This is a suitable approximation for modeling femtosecond pulse propagation in gases at standard conditions, which exhibit negligible linear losses [8]. The latter approximation will be used throughout this work, unless otherwise stated. It is furthermore assumed that the nonlinear response is isotropic and homogeneous. In combination with the paraxiality assumption $\vec{\nabla} \cdot \vec{E} \approx 0$, this leads to a decoupling of the vectorial components $\vec{E} = (E_x, E_y, E_z)$ in the propagation Eq. (2.9). Assuming linear polarization of the initial laser field, $\vec{E} = (E_x, 0, 0)$, the polarization is then preserved during beam propagation in the paraxial regime, and throughout the thesis, it is justified to switch to a scalar description, setting

$$\vec{E} = E\vec{e}_x, \qquad \vec{P}_{NL} = P_{NL}\vec{e}_x, \qquad \vec{J} = J\vec{e}_x \tag{2.10}$$

with orthogonal unit vectors $\vec{e}_x, \vec{e}_y, \vec{e}_z$. However, it should be noted that for large nonparaxiality, the latter assumptions cannot be maintained, leading to a nonlinear coupling of differently polarized states, as has recently been demonstrated in Ref. [16].

Although the second order wave equation (2.9) provides a convenient simplification of the full model Eq. (2.2), both the paraxiality criterion and the condition Eq. (2.8) have not been fully exploited yet. In fact, as demonstrated in [2, 3, 17], the second order wave equation can be factorized to yield a first order differential equation in z, a fact that greatly simplifies numerical beam propagation. A detailed derivation of this factorization procedure can be found in Refs. [2, 3]. Here, the method is outlined by means of the one-dimensional Helmholtz equation with an inhomogeneity h,

$$\frac{\partial^2 \widehat{E}}{\partial z^2} + k^2 \widehat{E} = \widehat{h} \tag{2.11}$$

where $k = n(\omega)\omega/c$ and $\widehat{E}(z, \omega)$ denotes the frequency domain representation of the electric field $E(z, t)$ in the time domain.

Fourier transform w.r.t z, $\widehat{E}(z, \omega) \rightarrow \widehat{E}_\beta(\beta, \omega)$, where β denotes the conjugate variable yields the equation

$$\widehat{E}_\beta = \frac{\widehat{h}_\beta}{k^2 - \beta^2}, \tag{2.12}$$

where it was used that $\widehat{\partial/\partial z} = -i\beta$ and the equation was formally solved for $\widehat{E}_\beta$. The rather formal manipulations leading to Eq. (2.12) can be substantiated by noting that the Fourier transform w.r.t. β,

$$G_\omega(z, z') = \int d\beta \frac{e^{-i\beta(z-z')}}{k^2(\omega) - \beta^2}, \tag{2.13}$$

corresponds to the Green's function $G(z, z')$ of the one-dimensional Helmholtz equation. This allows the construction of a solution to the inhomogeneous Eq. (2.11) according to

$$\widehat{E}(z, \omega) = \int dz' G_\omega(z, z') \widehat{h}(z', \omega). \tag{2.14}$$

However, note that appropriate boundary conditions [2] have to be supplied to solve the problem (2.11) using Eqs. (2.12) and (2.13).

Factorization of the Helmholtz equation is achieved by noting that Eq. (2.12) can be decomposed according to [3]

$$\widehat{E}_\beta \equiv \frac{\widehat{h}_\beta}{\beta^2 - k^2} = \widehat{E}_\beta^+ + \widehat{E}_\beta^-, \tag{2.15}$$

where forward and backward propagating electric field components $\widehat{E}_\beta^\pm$ were defined according to

$$\widehat{E}_\beta^+ = -\frac{\widehat{h}_\beta}{2k} \frac{1}{\beta + k}, \quad \widehat{E}_\beta^- = \frac{\widehat{h}_\beta}{2k} \frac{1}{\beta - k}. \tag{2.16}$$

The Helmholtz equation in the z-domain is therefore equivalent to the set of first-oder differential equations

$$(\partial_z + ik)\widehat{E}^+ = \frac{\widehat{h}}{2k}, \ (\partial_z - ik)\widehat{E}^- = \frac{\widehat{h}}{2k} \tag{2.17}$$

The wave fields $E^\pm$ correspond to waveforms traveling into the positive and negative z directions. In the linear regime, they evolve independently. The inhomogeneous three-dimensional Helmholtz Equation (2.9) allows a completely analogous factorization, with the subtle difference that the inhomogeneity h may depend on the field E to model pulse propagation in the nonlinear regime. In this case, the factorized Helmholtz equations for the forward- and backward propagating field components are nonlinearly coupled. However, it is shown in Ref. [3] that for an initial field $E = E^+ + E^-$ with a dominant forward-propagating field component E^+, the backward-propagating component E^- stays small along z-propagation and can be neglected, as long as the paraxiality criterion $k_\perp / |\vec{k}| \ll 1$ and the condition (2.8) are fulfilled.

As shown in Sect. 2.3, these criteria are usually satisfied in filamentary propagation, which justifies the assumption $\widehat{E} = \widehat{E}^+$. The factorization procedure thus yields a first order partial differential equation for the forward-propagating field,

$$\frac{\partial \widehat{E}}{\partial z} = \frac{i}{2k(\omega)} \nabla_\perp^2 \widehat{E} + ik(\omega)\widehat{E} + \frac{i\mu_0\omega^2}{2k(\omega)} \left(\widehat{P_{\mathrm{NL}}} + i\frac{\widehat{J}}{\omega} \right). \tag{2.18}$$

This equation has originally been used in Ref. [4] as a starting point to analyze supercontinuum generation in photonic crystal fibers. While Eq. (2.18) describes freely propagating pulses in a nonlinear medium, a rigorous derivation of an equation analogous to the FME, describing forward-propagting pulses in a guided geometry, has recently been given in Refs. [18, 19].

2.2 The Nonlinear Optical Response

This section is devoted to the nonlinear response of the material to the intense laser field. The basic assumption of perturbative nonlinear optics is that the nonlinear polarization P_{NL} of an isotropic medium can be decomposed as

$$P_{\mathrm{NL}} = P^{(3)} + P^{(5)} + P^{(7)} + \cdots . \tag{2.19}$$

As only isotropic, centrosymmetric media are examined in the following, all even-order contributions $P^{(2k)}$ vanish identically [20]. Demanding that the nonlinear response respects time-translational invariance of the dynamical equation leads to the following expression for the n-th order contribution in the time domain[2] [22].

$$P^{(n)}(\vec{r}, t) = \epsilon_0 \int_{-\infty}^{\infty} d\tau_1 \int_{-\infty}^{\infty} d\tau_2 ... \int_{-\infty}^{\infty} d\tau_n R^{(n)}(\tau_1, \tau_2, ..., \tau_n)$$
$$\times E(\vec{r}, t - \tau_1) E(\vec{r}, t - \tau_2)...E_n(\vec{r}, t - \tau_n). \tag{2.20}$$

In the frequency domain, this translates into

$$P^{(n)}(\vec{r}, \omega) = \epsilon_0 \int \cdots \int \chi^{(n)}(-\omega_\sigma; \omega_1, ..., \omega_n) E(\vec{r}, \omega_1)...E(\vec{r}, \omega_n)\delta(\omega - \omega_\sigma)d\omega_1...d\omega_n, \tag{2.21}$$

where $\omega_\sigma = \omega_1 + \omega_2 + \cdots + \omega_n$, and only homogeneous media are considered for which the response kernel $R^{(n)}$ and the susceptibilities $\chi^{(n)}$ are independent of position. The n-th-order contribution to the nonlinear polarization is frequently considered as resulting from an $n + 1$-photon process interacting with bound electronic states. From this point of view, the delta function in the integrand ensures conservation of photon energy, $\hbar\omega = \hbar\omega_1 + \cdots + \hbar\omega_n$.

---

[2] As in the case of the linear polarization, spatial dispersion modeled by a wave-vector dependent nonlinear susceptibility $\chi^{(n)}(\omega_1, \cdots, \omega_n, \vec{k}_1, \cdots, \vec{k}_n)$ was disregarded. Spatially dispersive nonlinearities involve a nonlocal optical response and can arise from thermal effects or may occur in dipolar Bose-Einstein condensates [21].

2.2.1 Third-Order Response to a Monochromatic Wave

In the following, the impact of the first non-vanishing order $P^{(3)}$ on a monochromatic plane wave of frequency ω_0 and amplitude E_0 propagating into the positive z-direction with wave-vector $k_0 \equiv k(\omega_0) = n(\omega_0)\omega_0/c$,

$$E(\vec{r}, t) = E_0 \cos(\omega_0 t + kz + \varphi) \tag{2.22}$$

will be discussed. With Euler's formula for the cosine, this may be decomposed according to

$$E(\vec{r}, t) = \frac{1}{2}\left(\mathcal{A}e^{i\omega_0 t + ik_0 z} + \mathcal{A}^* e^{-i\omega_0 t - ik_0 z} \right), \tag{2.23}$$

where

$$\mathcal{A} = E_0 e^{i\varphi}. \tag{2.24}$$

With the help of Eq. (2.23), the frequency-domain representation Eq. (2.21) of the third-order nonlinear polarization induced by a monochromatic plane wave may be written as [20]

$$
\begin{aligned}
P^{(3)}(\omega) &= \frac{3}{8}\epsilon_0 \chi^{(3)}(-\omega_0; \omega_0, \omega_0, -\omega_0)|\mathcal{A}|^2 \mathcal{A}\delta(\omega - \omega_0)e^{ikz} \\
&+ \frac{3}{8}\epsilon_0 \chi^{(3)}(\omega_0; -\omega_0, -\omega_0, \omega_0)|\mathcal{A}|^2 \mathcal{A}^*\delta(\omega + \omega_0)e^{-ikz} \\
&+ \frac{1}{8}\epsilon_0 \chi^{(3)}(-3\omega_0; \omega_0, \omega_0, \omega_0)\mathcal{A}^3\delta(\omega - 3\omega_0)e^{i3kz} \\
&+ \frac{1}{8}\epsilon_0 \chi^{(3)}(3\omega_0; -\omega_0, -\omega_0, -\omega_0)\mathcal{A}^{*3}\delta(\omega + 3\omega_0)e^{-i3kz}.
\end{aligned}
\tag{2.25}
$$

It follows that the polarization $P^{(3)}$ oscillates at frequencies $\pm 3\omega_0$ and $\pm\omega_0$. While the latter give rise to a nonlinear refractive index change as will be detailed below, the former correspond to the generation of a third-harmonic wave copropagating with the fundamental wave, a phenomenon known as third-harmonic generation (THG). However, the expression (2.25) shows that there exists a mismatch between the wave-vector $3k(\omega_0)$ of the polarization and the wave-vector $k(3\omega_0)$ of the radiated harmonic wave, $\Delta k = k(3\omega_0) - 3k(\omega_0)$, whenever the medium exhibits nontrivial dispersion $n(3\omega_0) \neq n(\omega_0)$[20]. In general, this will lead to destructive interference of the third harmonic waves generated at different positions unless suitable phase-matching techniques [20] are applied which ensure vanishing of the wave-vector mismatch Δk. Harmonic generation is therefore disregarded in the following, focusing the attention to self-induced refractive index changes.

2.2.2 Third-Order Response to an Optical Pulse

While Eq. (2.25) was derived for a monochromatic plane wave, filamentation is only observed for sufficiently high peak powers of the laser pulse of the order of 100 GW, which is impossible to achieve with monochromatic light. Instead, only pulsed laser sources generating ultrashort pulses with durations of the order of some ten femtoseconds are capable of providing the required peak optical powers. Equation (2.25) therefore has to be generalized for ultrashort optical pulses. The subsequent discussion is greatly simplified by introducing so-called complex-valued analytic signals. With the decomposition (2.23), the real-valued monochromatic wave is seen to consist of positive and negative frequency components. This can be generalized for arbitrary time-dependence of the electric field, using that the Fourier transform of any real-valued function $F(t)$ satisfies $\widehat{F}(-\omega) = \widehat{F}^*(\omega)$. This reveals that the information contained in the negative frequency components of F can be considered redundant, and instead of the real-valued electric field E, the so called analytic signal E_A [19] is considered in the following. This is composed of the positive frequency components of E according to

$$E_A(\vec{r}, t) = 2 \int_0^\infty d\omega \widehat{E}(\vec{r}, \omega) e^{-i\omega t}. \tag{2.26}$$

From this, the electric field may easily be reconstructed according to

$$E(\vec{r}, t) = \frac{1}{2}(E_A(\vec{r}, t) + E_A^*(\vec{r}, t)). \tag{2.27}$$

It is moreover useful to factorize the analytic signal E_A into an envelope $\mathcal{A}$ and an exponential oscillating at the carrier-frequency ω_0 of the laser field,

$$E_A(\vec{r}, t) = \mathcal{A}(\vec{r}, t) e^{-i\omega_0 t}, \tag{2.28}$$

where the carrier frequency ω_0 denotes the mean frequency [23]

$$\omega_0 = \frac{\int_{-\infty}^{\infty} d\omega |\widehat{E}|^2 \omega}{\int_{-\infty}^{\infty} d\omega |\widehat{E}|^2}. \tag{2.29}$$

In the frequency domain, the definition (2.28) corresponds to the identity $\widehat{\mathcal{A}}(\vec{r}, \omega) = \widehat{E}_A(\vec{r}, \omega + \omega_0)$, which shows that $\mathcal{A}$ has zero mean frequency, corresponding to the removal of the fast carrier oscillations at ω_0, leaving only a pulse envelope. In what follows, the generalization of Eq. (2.25) for short laser pulses shall be discussed. However, experimental or theoretical data describing the dispersion of $\chi^{(3)}$ over a large frequency range often vary by orders of magnitude [24]. More reliable data is available from measurements or calculations of $\chi^{(3)}$ at a single frequency. Consequently, it is henceforth assumed that the spectral bandwidth of the pulse is small with respect to the frequency scale on which $\chi^{(3)}$ shows notable variation. Then, it is possible to show [8] that the third-order polarization induced by the electromagnetic pulse is given by

$$P^{(3)}(\vec{r},t) = \frac{3}{8}\epsilon_0\chi^{(3)}(-\omega_0;\omega_0,\omega_0,-\omega_0)|\mathcal{A}(\vec{r},t)|^2\mathcal{A}(\vec{r},t)e^{-i\omega_0 t} + c.c.$$

$$+ \frac{1}{8}\epsilon_0\chi^{(3)}(-3\omega_0;\omega_0,\omega_0,\omega_0)\mathcal{A}^3(\vec{r},t)e^{-i3\omega_0 t} + c.c.. \tag{2.30}$$

Neglecting again the THG term oscillating at $3\omega_0$, the third order polarization gives rise to an intensity dependent change of the refractive index. This is due to the fact that sufficiently strong electromagnetic fields can distort the electronic distribution within in the medium, which gives rise to a modified refractive index. This effect is also referred to as the all-optical Kerr effect [25] and should not be confused with the electro-optic (DC) Kerr effect [26], where a static electric field induces birefringence in the material. In order to further evaluate the third order contribution to the intensity dependent refractive index (IDRI), it is useful to introduce the optical intensity I. As the energy density of an electric field is proportional to the square of the electric field strength, it follows that the optical intensity is given by [23]

$$I(\vec{r},t) = \epsilon_0 c n_0 \frac{1}{T} \int\limits_{t-T/2}^{t+T/2} E^2(\vec{r},t')dt', \tag{2.31}$$

where $n_0 \equiv n(\omega_0)$ denotes the refractive index at the center frequency, and the average over one optical cycle of duration $T = 2\pi/\omega_0$ was taken. Demanding that the envelope $\mathcal{A}$ defined in Eq. (2.28) varies slowly compared to the carrier oscillation at ω_0, it follows that the above relation for the cycle-averaged intensity can be evaluated to give

$$I = \frac{1}{2}n_0\epsilon_0 c|\mathcal{A}|^2 \tag{2.32}$$

Including only the third-order nonlinear polarization, it can be deduced from Eq. (2.30) that the IDRI due to the all-optical Kerr effect is given by

$$n(I) = n_0 + n_2 I, \tag{2.33}$$

where n_2 denotes the second order nonlinear refractive index which is given by

$$n_2 = \frac{3}{4n_0^2\epsilon_0 c}\chi^{(3)} \tag{2.34}$$

Note that for the latter derivation, both linear and nonlinear absorption were disregarded, which allows to impose $\mathrm{Im}\chi^{(1)} = \mathrm{Im}\chi^{(3)} = 0$. In fact, this approximation is frequently justified in the context of femtosecond filamentation [8, 12]. A more detailed discussion of higher-order nonlinear refraction and absorption coefficients and their relation to the nonlinear susceptibilities $\chi^{(n)}$ is provided in Sect. 2.3. Indeed, it is one of the main conclusions both, of recent experimental results [27–29], and of the theoretical investigations in Chap. 4, that higher-order nonlinearities $\chi^{(n)}$ for $n > 3$ actually play a greater role than previously supposed.

2.2.3 Plasma Response

Besides the all-optical Kerr effect, an important contribution to the nonlinear refractive index is given by free carriers. In fact, the intensities achieved within femtosecond filaments are sufficiently high to trigger photoionization processes. The femtosecond laser pulse thus propagates in a self-generated plasma. The current density J taking into account the generation of free carrier by photoionization can be decomposed according to

$$J = J_{\mathrm{FC}} + J_{\mathrm{PI}}, \tag{2.35}$$

where J_{FC} is the current density of free carriers subject to the electric field E, while J_{PI} accounts for losses due to photoionization. Both quantities couple to the FME Eq. (2.18). The dynamics of the free carriers is treated in terms of the Drude model [30, 31] according to

$$\frac{\partial J_{\mathrm{FC}}}{\partial t} + \frac{J_{\mathrm{FC}}}{\tau_c} = \frac{q_e^2 \rho}{m_e} E \tag{2.36}$$

Here, q_e and m_e denote electron charge and mass, respectively, ρ denotes the number of free carriers per unit volume and τ_c represents the mean time between collision of free carriers. In the frequency domain, Eq. (2.36) can be formally solved for the Fourier transform $\widehat{J}_{\mathrm{FC}}$, and it is found that the current of free carriers (2.18) is given by [8]

$$-\frac{\mu_0 \omega}{2k(\omega)} \widehat{J}_{\mathrm{FC}} = \frac{1}{2k(\omega)}\left(-\frac{\omega n_0 \sigma(\omega)}{c} - i\frac{\omega_0^2}{c^2 \rho_c (1 + \nu_e^2/\omega^2)} \right)\widehat{\rho E} \tag{2.37}$$

where $n_0 = n(\omega_0)$ is the refractive index at the carrier frequency, $\nu_e = 1/\tau_c$ and $\rho_c = \omega_0^2 m_e \epsilon_0 / q_e^2$ is the critical density of free carriers for which the plasma becomes opaque for a laser beam of carrier frequency ω_0. The cross-section for collision of free carriers is given by

$$\sigma(\omega) = \frac{q_e^2}{m_e \epsilon_0 n_0 c \nu_e (1 + \omega^2/\nu_e^2)}. \tag{2.38}$$

In Eq. (2.37), the loss term involving the cross-section $\sigma(\omega)$ accounts for collisional ionization by free carriers accelerated in the laser field. As this process consumes electromagnetic energy, it is frequently referred to as inverse Bremsstrahlung. In contrast, the term involving the purely imaginary prefactor of $\widehat{\rho E}$ corresponds to the change of the refractive index due to the plasma and will be discussed below.

In addition, direct photoionization of neutral atoms takes energy from the laser field. This requires the introduction of the loss current [8],

$$ J_{\mathrm{PI}} = \frac{k_0}{\omega_0 \mu_0} \frac{U_i w(I)}{I} (\rho_0 - \rho) E. \tag{2.39} $$

This quantity depends on the ionization potential U_i of the gas species and on the neutral density ρ_0. Furthermore, it depends on the ionization rate $w(I)$. A theoretical derivation of the ionization rate of atoms or molecules subject to intense laser fields has been performed by several independent researchers [10, 32–36]. Throughout this work, the results of Perelomov, Popov and Terent'ev (PPT) [32, 35] are applied. A deeper discussion of the PPT model is presented in Chap. 4 of this thesis. As the ionization depends highly nonlinear on the intensity, it is justified to assume that only frequency components of the pulse close to the carrier frequency ω_0 contribute to ionization processes. Therefore, for the collisional cross section the replacement $\sigma(\omega) \rightarrow \sigma(\omega_0)$ is performed throughout. It then follows that the density ρ of the self-generated plasma satisfies the rate equation

$$ \frac{\partial \rho}{\partial t} = w(I)(\rho_0 - \rho) + \frac{\sigma(\omega_0)}{U_i} \rho I. \tag{2.40} $$

While typical timescales relevant in filamentation are of the order of $10^{-13} - 10^{-14}$ s, recombination of ions and electrons takes place on a nanosecond timescale. This justifies to neglect recombination effects in Eq. (2.40). The first term on the r.h.s. of Eq. (2.40) accounts for photoionization, while the second term models the contribution of collisional ionization to the electron density.

2.3 Contributions to the Nonlinear Refractive Index

2.3.1 Plasma Contributions

For the case of a monochromatic plane wave of frequency $\omega = \omega_0$, leading to $k(\omega) = k(\omega_0) = k_0$ and $n(\omega) = n(\omega_0) = n_0$, the FME Eq. (2.18) reduces to

$$ \frac{\partial \widehat{E}}{\partial z} = -i \frac{\omega_0}{c} (n_0 + \Delta n_p) \widehat{E}, \tag{2.41} $$

where additionally, losses due to collisional ionization, i.e., $\nu_e \to 0$, and the nonlinear polarization were neglected. This shows that for $\nu_e = 0$, the contribution of the free carriers to the refractive index is given by $\Delta n_p = -\rho/2n_0^2\rho_c$. In contrast, using the Drude model of a collisionless plasma and the wave Eq. (2.9), it turns out that the presence of plasma in a medium with neutral refractive index n_0 lowers the refractive index according to [37]

$$n = \sqrt{n_0^2 - \frac{\omega_p^2}{\omega^2}}, \tag{2.42}$$

where $\omega_p = \sqrt{\rho q_e^2/m_e\epsilon_0} = \omega_0\sqrt{\rho/\rho_c}$ is the plasma frequency. The obvious discrepancy arises from the approximations introduced with the FME: the term accounting for the linear polarization $\sim k^2(\omega)\widehat{E}$ exhibits a quadratic dependence on the wave vector k, while the current density J_{FC} enters linearly. In contrast, due to the factorization procedure, the linear polarization gives rise to a term $\sim k(\omega)\widehat{E}$ on the r.h.s. of Eq. (2.9), while the term containing the current is not affected by the factorization and enters linearly. However, for $\rho \ll \rho_c$, Eq. (2.42) may be approximated according to

$$n = n_0 - \frac{\rho}{2n_0^2\rho_c}, \tag{2.43}$$

which corresponds to the plasma induced refractive index change derived from the FME. Thus, the inequality $\rho \ll \rho_c$ provides an additional criterion for the validity of the FME. As an example of practical relevance, filamentary propagation of a pulsed femtosecond laser beam emitted by a Ti:sapphire amplifier (with a center wavelength of 800 nm) is considered. Assuming that the pulse is propagating in a gaseous medium of atmospheric pressure, the ratio of ionized particles is of the order of 10^{-3} [38], i.e. $\rho \approx 3 \times 10^{16}$ cm^{-3}, while the critical plasma density for the given wavelength is $\rho_c \approx 2 \times 10^{21}$ cm^{-3}. Under these assumptions the plasma-induced refractive index change in femtosecond filaments is therefore of the order of $\rho/\rho_c \approx 10^{-5}$, which justifies the approximations introduced with the FME Eq. (2.18).

2.3.2 Contributions Due to the All-Optical Kerr Effect

In linear optics, the refractive index n_0 and absorption coefficient α_0 are related to the complex dielectric permittivity ϵ according to

$$(n_0 + i\alpha_0 c/2\omega)^2 = \epsilon. \tag{2.44}$$

Using

$$\widehat{D} \equiv \epsilon_0\epsilon\widehat{E} = \epsilon_0\widehat{E} + \widehat{P}^{(1)} \tag{2.45}$$

satisfied by the dielectric displacement and Eq. (2.7) for the linear polarization $\widehat{P}^{(1)}$ it follows that $\epsilon(\omega) = 1 + \chi^{(1)}(\omega)$. This consideration can be generalized to the case of nonlinear optics if it is assumed that the spectral bandwidth of the optical pulse E is small compared to the frequency scale on which the nonlinear susceptibilities $\chi^{(n)}$ show notable dispersion. In analogy to the reasoning that led to Eq. (2.30) for the third-order susceptibility $\chi^{(3)}$, an envelope description (Eq. 2.28) is introduced to identify the self-refraction terms contributing to the nonlinear polarization $P^{(n)}$. This yields an intensity dependent dielectric permittivity [39]

$$\epsilon(I) = 1 + \chi^{(1)}(\omega_0) + \sum_{k \geq 1} C^{(k)} \chi_{\omega_0}^{(2k+1)} |\mathcal{A}|^{2k}, \tag{2.46}$$

where the intensity I is related to the envelope $\mathcal{A}$ according to Eq. 2.32. The factor $C^{(k)}$ follows from combinatorial considerations [40] and is given by

$$C^{(k)} = \frac{(2k+1)!}{2^{2k} k! (k+1)!}, \tag{2.47}$$

and $\chi_{\omega_0}^{(n)}$ denotes the nth-order nonlinear susceptibility associated to self-refraction, e.g., for the third order polarization, $\chi_{\omega_0}^{(3)} = \chi^{(3)}(-\omega_0, \omega_0, \omega_0, -\omega_0)$, while $\chi^{(1)}(\omega_0)$ denotes the linear susceptibility at frequency ω_0. From Eq. (2.46), a nonlinear refractive index $n(I)$ and a nonlinear absorption coefficient $\alpha(I)$ can be defined by generalizing Eq. (2.44) according to

$$(n(I) + i\alpha(I)c/2\omega)^2 = \epsilon(I). \tag{2.48}$$

Compact approximate expressions for $n(I)$ and $\alpha(I)$ can be derived if it is assumed that the nonlinear refraction and absorption changes $\Delta n(I) = n(I) - n_0$ and $\Delta \alpha(I) = \alpha(I) - \alpha_0$ are sufficiently small such that only first order contributions of these quantities have to be considered. In addition, it is assumed that the linear absorption coefficient α_0 satisfies $\alpha c/\omega \ll n_0$ [22], which leads to the following expressions for the nonlinear refractive index and absorption coefficient,

$$\begin{aligned} n(I) &= n_0 + \sum_{k \geq 1} n_{2k} I^k \\ \alpha(I) &= \alpha_0 + \sum_{K \geq 2} \beta_K I^{K-1}. \end{aligned} \tag{2.49}$$

The coefficients n_{2k} and β_K are related to the real and imaginary part of the nonlinear susceptibilities $\chi^{(2k+1)}$ pursuant to

$$n_{2k} = \frac{2^{k-1} C^{(k)}}{n_0 (n_0 \epsilon_0 c)^k} \mathrm{Re} \chi^{2k+1} \tag{2.50}$$

$$\beta_K = \frac{\omega_0}{c} \frac{2^{K-1} C^{(K-1)}}{n_0 (n_0 \epsilon_0 c)^{K-1}} \mathrm{Im} \chi^{2K-1}. \tag{2.51}$$

It is interesting to note that the approximations involved in defining a nonlinear refractive index are closely related to the approximation (2.8) made during the derivation of the FME. In fact, in terms of refractive index changes, the condition on P_{NL} translates itself into $\Delta n(I) \ll n_0$.

In Chap. 4 it is shown that the nonlinearly induced refractive index changes Δn are small for the noble gases helium, neon, argon, krypton and xenon, which are the most commonly used media in experimental femtosecond filamentation. For these gases, it can be shown that Δn varies between $\sim 10^{-5}$ and $\sim 10^{-7}$ for intensities up to $40\,\mathrm{TW/cm^2}$ (xenon) and $300\,\mathrm{TW/cm^2}$ (helium), respectively. As the error introduced by approximation (2.49) is of the order Δn^2, the use of the FME is clearly justified.

2.4 An Envelope Equation for Few-Cycle Optical Pulses

A further simplification of the FME may be obtained by imposing certain restrictions on the envelope $\mathcal{A}$. Besides assuming that the envelope varies slowly in time, it has to be imposed that the envelope varies slowly in the spatial coordinate z. Thus, for the following, besides subtracting the carrier oscillations at ω_0 in time, a subtraction of the spatial oscillations along the propagation direction z is necessary. These oscillations are governed by the z-component k_z of the wave-vector. However, assuming paraxial propagation, it is found that $k_\perp/k \ll 1$ which is equivalent to $k_z \approx k_0$. The electric field is then rewritten in terms of amplitudes that are slowly varying both in time and space pursuant to

$$E(\vec{r}, t) = \sqrt{c_1} \left(\mathcal{E}(\vec{r}, t) e^{ik_0 z - i\omega_0 t} + \mathcal{E}^*(\vec{r}, t) e^{-ik_0 z + i\omega_0 t} \right). \tag{2.52}$$

The normalization factor $c_1 = \mu_0/(n_0^2 \epsilon_0)$ is chosen such that $I = |\mathcal{E}|^2$. The envelopes $\mathcal{E}$ and $\mathcal{A}$ are related by $\mathcal{E} = \mathcal{A} \exp(-ik_0 z)/2\sqrt{c_1}$. The requirements leading to an envelope $\mathcal{E}$ varying slowly both in t and z then read

$$\left| \frac{\partial}{\partial z} \mathcal{E} \right| \ll k_0 |\mathcal{E}|, \tag{2.53}$$

and

$$\left| \frac{\partial}{\partial t} \mathcal{E} \right| \ll \omega_0 |\mathcal{E}|. \tag{2.54}$$

These restrictions provide the slowly varying envelope approximation (SVEA) [41]. With these conditions, a simple first order PDE in z for the envelope $\mathcal{E}$ was obtained [5], which, neglecting plasma response and nonlinearities higher than third order, corresponds to the Nonlinear Schrödinger Equation, see Appendix A for a brief mathematical introduction. This equation has successfully been applied to explain various phenomena during the early days of nonlinear optics. However, the latter equation fails to correctly describe ultra-broadband pulses as they arise for example in femtosecond filaments. This is due to the fact that for ultra-broadband pulses, the slowly varying envelope ceases to be a meaningful concept, especially for pulses consisting only of a few-cycles of the optical carrier field. However, a generalized envelope equation capable of describing the propagation of few-cycle pulses can be derived from the FME, which yields [6]

$$\partial_z \mathcal{E} = \frac{i}{2k_0} T^{-1} \Delta_\perp \mathcal{E} + i \mathcal{D} \mathcal{E} + i \frac{\omega_0}{c} n_2 T |\mathcal{E}|^2 \mathcal{E} - i \frac{k_0}{2\rho_c} T^{-1} \rho(I) \mathcal{E}$$
$$- \frac{\sigma}{2} \rho \mathcal{E} - \frac{U_i w(I)(\rho_{nt} - \rho)}{2I} \mathcal{E}, \tag{2.55}$$

$$\partial_t \rho = w(I)(\rho_{nt} - \rho) + \frac{\sigma}{U_i} \rho I. \tag{2.56}$$

In the above equation also referred to as the nonlinear envelope equation (NEE), only the third-order nonlinearity $\chi^{(3)}$ was taken into account. Furthermore, a transformation of variables $t \rightarrow t - z/v_g(\omega_0)$ to a frame comoving with the group velocity $v_g(\omega) = (dk(\omega)/d\omega)^{-1}$ of the laser pulse was performed, and it was used that $v_g \approx c$ for gaseous media at standard conditions. The operator T ensures validity of the model in the few-cycle domain and is given by

$$T = 1 + \frac{i}{\omega_0} \partial_t. \tag{2.57}$$

while the operator $\mathcal{D}$ is given, in the frequency domain, by

$$\widehat{\mathcal{D}}(\omega) = k(\omega) - k_0 - (\omega - \omega_0) \left.\frac{\partial k}{\partial \omega}\right|_{\omega = \omega_0} \tag{2.58}$$

A Taylor expansion of this expression followed by a Fourier transform, yields the following expression for the operator $\mathcal{D}$ in the time domain,

$$\mathcal{D} = \frac{1}{2!} \beta_2 \frac{\partial^2}{\partial t^2} + \frac{1}{3!} \beta_3 \frac{\partial^3}{\partial t^3} + \cdots \tag{2.59}$$

where $\beta_n = d^n k/d\omega^n$, evaluated at $\omega = \omega_0$, and it was used that the Fourier transform $\mathcal{F}$ satisfies

$$\mathcal{F}^{-1}(\omega \widehat{G}(\omega)) = i \frac{d}{dt} G(t) \tag{2.60}$$

The operator $\mathcal{D}$ describes the dispersion of the temporal pulse profile of the pulse due to the fact that different frequency components of the pulse propagate with different velocities. For a narrow-band pulse, it often suffices to employ the power series representation Eq. (2.59) up to some finite order. However, for ultra-broadband pulses, it is more appropriate to evaluate $\mathcal{D}(\omega)$ by using Sellmeier type equations for $n(\omega)$ [41, 42]. Alternatively, in the context of fiber optics, a recent approach involves approximation of $\mathcal{D}(\omega)$ using rational functions [43]. The frequency domain representation $\widehat{T}$ of T reads

$$\widehat{T} = 1 + \frac{\omega}{\omega_0} \tag{2.61}$$

This operator emerges by setting $k(\omega) \approx n_0\omega/c$ on the r.h.s. of the FME (2.18), where it is assumed that the carrier wavelength is far from a medium resonance such that $n(\omega)$ exhibits weak dispersion. Subsequently, in the ω dependent prefactors of $\widehat{E}$, $\widehat{P}_{\mathrm{NL}}$ and $\widehat{J}$, the identity $\omega = \omega_0(1 + (\omega - \omega_0)/\omega_0)$ is employed. The operator T then emerges by noting that $\omega - \omega_0 \to i\partial_t$, where ∂_t is restricted to act on the **pulse envelope** $\mathcal{E}$. With these approximations, the envelope Eq. (2.55) is obtained after performing the Fourier transform into the time domain. It suitably models the propagation of few-cycle pulses given that the electric field E satisfies

$$\left| \frac{\partial E}{\partial z} \right| \ll k_0 |E|, \tag{2.62}$$

This holds when the electric field evolves slowly during z-propagation. Therefore, the approximations leading to the NEE are referred to as the Slowly Evolving Wave Approximation (SEWA). Note that the SVEA corresponds to setting $T = 1$ in the NEE (2.55). In contrast to the SVEA, the SEWA does not imply a limitation of the pulse duration and can be used to model the propagation of few-cycle pulses in media subject to the additional limitation

$$\left| k_0 - \frac{\omega_0}{v_g(\omega_0)} \right| / k_0 \ll 1, \tag{2.63}$$

demanding that the group velocity differs from the phase velocity only marginally.

Historically, the generalized envelope equation was derived by Brabec and Krausz prior to the FME. However, as shown above, the envelope Eq. (2.55) can be derived from the more general FME in a rather straightforward manner. When wave-mixing phenomena like THG or sum frequency generation in filaments are investigated, the FME should clearly be preferred against the envelope description [44]. Nevertheless, only self-refraction effects are considered throughout this thesis, as the radiation due to THG is typically poorly phase-matched in femtosecond filaments and can be neglected. Further on, like the FME, the NEE provides a valid model for propagation of few-cycle pulses as shown in the above-cited references.

Therefore, the Eqs. (2.55) and (2.56) suitably describe femtosecond filamentation for the medium and input pulse parameters considered here.

2.4.1 Reduction to the Cylindrically Symmetric Case

It turns out that the complexity of solving Eq. (2.55) numerically can be strongly reduced by imposing an additional symmetry constraint, i.e. cylindrical symmetry along the propagation direction z. As discussed in Sect. 2.5, the evolution Eq. (2.55) is subject to an azimuthal modulation instability which amplifies small amplitude, radially asymmetric perturbations on a cylindrically symmetric beam, leading to a loss of cylindrical symmetry and eventually a multiple spatial break-up of the beam also known as multifilamentation. However, it follows from the experimental observations of Ref. [45] that the radial symmetry of the input beam is preserved during filamentary propagation for input powers not exceeding roughly $5 - 6P_{\mathrm{cr}}$ [45], where P_{cr} is the critical power for self-focusing, as discussed in detail in Sect. 2.5. In the latter experimental work, 45 fs, 5 mJ-pulses emitted by a regenerative Ti:sapphire amplifier were focused into a 1.5 m long cylindrical gas cell filled with argon. For the chosen input beam parameters, the above constraint on the input power translates itself into a constraint for the pressure within the argon gas cell to values below 60 kPa, thereby limiting the nonlinearity n_2. Alternatively, the energy of the input pulse can be diminished using an adjustable diaphragm. Indeed, a carefully adjusted diaphragm has frequently been proven to be a suitable measure to avoid multiple filamentation [46]. Under these assumptions, it suffices to consider cylindrically symmetric solutions $\mathcal{E}(r, z, t)$ of Eq. (2.55). In this case the Laplacian in Eq. (2.55) can be reduced to its radial component

$$\Delta_\perp = (1/r)\partial_r r \partial_r, \tag{2.64}$$

where the radial coordinate is given by $r = \sqrt{x^2 + y^2}$.

2.5 Properties of Filamentary Propagation

In the following, various limiting cases of the envelope Eq. (2.55) are considered, and a discussion of the phenomena relevant in the respective regimes is provided.

2.5.1 Dispersion

Dispersion is usually referred to as the frequency dependence of certain material properties governing the response to an external optical field, such as the refractive index $n = n(\omega)$ or the nonlinear susceptibilities [14]. In linear optics, an external optical field induces a frequency dependent polarization, which may reshape the irradiated optical pulse during propagation, as different frequency components of the pulse propagate with unequal phase velocities in the medium. Neglecting the

nonlinear response terms and assuming that the pulse is sufficiently long such that
the SVEA can be applied, it is a reasonable approximation to set $T = 1$ in Eq. (2.55).
In addition, only plane waves propagating into the positive z-direction are considered.
Then, it follows that the NEE reduces to

$$\partial_z \mathcal{E} = i \mathcal{D} \mathcal{E}. \tag{2.65}$$

The dispersion operator accounts for group-velocity dispersion (GVD) governed
by β_2, third-order dispersion (TOD) governed by β_3 and higher-order dispersion
terms [41]. Here, the effect of GVD on an initially Gaussian-shaped temporal profile,

$$\mathcal{E}(0, t) = N(0)e^{-t^2/t_p(0)^2} \tag{2.66}$$

will be discussed. Neglecting all higher order terms but GVD, Eq. (2.65) can be
integrated to yield [41]

$$\mathcal{E}(z, t) = N(z)e^{-t^2/t_p^2(z) - iC(z)t^2/t_p^2(z)} \tag{2.67}$$

with

$$N(z) = \frac{N(0)}{\sqrt{1 - i\frac{z}{L_D}}}, \quad t_p(z) = t_p(0)\sqrt{1 + (z/L_D)^2}, \quad C(z) = z/L_D, \tag{2.68}$$

where $L_D = t_p^2(0)/2|\beta_2|$. The expressions for the normalization constant $N(z)$ and
the pulse duration $t_p(z)$ show that the amplitude of the pulse decreases, while the
duration $t_p(z)$ increases along z. The chirp factor $C(z)$ shows that GVD introduces
a linear frequency chirp on the pulse. A discussion of chirped Gaussian pulses can
be found in the introduction to the subsequent chapter dealing with pulse compression techniques. A characteristic length-scale on which these processes take place is
provided by L_D.

While group velocity dispersion β_2 introduces a symmetric temporal broadening of the pulse envelope, the odd-order terms β_{2k+1} terms introduce an asymmetric temporal stretching. Especially, the impact of TOD on Gaussian pulses can be
described analytically in terms of Airy functions [41]. In filamentation, one typically
focuses pulsed femtosecond laser beam into noble gases or air at atmospheric pressure. However, these exhibit a relatively weak dispersion [42]. For example, argon
at standard conditions has $\beta_2 = 0.2\,\text{fs}^2/\text{cm}$, such that the characteristic length scale
$L_D = 62.5\,\text{m}$ for a 50 fs initial pulse. In fact, one may estimate that GVD only
becomes relevant for pulses with initial duration $<10\,\text{fs}$, for which L_D approaches
the typical propagation distances of the order of $\sim 1\,\text{m}$ used in experimental investigations of filamentary propagation.

2.5.2 Self-Phase Modulation

Self-phase modulation arises from the intensity dependence of the refractive index, $n = n_0 + n_2 I$. This nonlinear effect can result in a substantial spectral broadening of an optical pulse, leading to the formation of a white-light supercontinuum. In condensed media, this was first observed in Ref. [47].

In order to investigate its impact on the evolution of a laser pulse, it is assumed that the optical intensity is sufficiently low not to trigger photo-ionization. Further on, it is assumed that the SVEA can be applied, leading to the condition that the pulse has to be much longer than the optical cycle, such that setting $T = 1$ in Eq. (2.55) is justified. If one additionally assumes that the dispersion length L_D is large, it is possible to neglect dispersion, setting $\mathcal{D} = 0$. Using a polar decomposition of the complex envelope $\mathcal{E}(z, t)$ of a plane wave propagating in the z-direction according to $\mathcal{E}(z, t) = |\mathcal{E}(z, t)|e^{-i\phi(z,t)}$, it can be inferred from the dynamical Eq. (2.55) that the temporal phase $\phi(z, t)$ of the pulse acquires a self-induced temporal phase-shift according to [12, 20]

$$\phi(z + \Delta z, t) = \phi(z, t) - \frac{\omega_0}{c} n_2 |\mathcal{E}(z, t)|^2 \Delta z. \tag{2.69}$$

From this, the SPM induced change of the instantaneous frequency, calculated as the time derivative of the instantaneous phase $\phi(t)$, is given by

$$\Delta\omega(t) = -\frac{\omega_0}{c} n_2 \frac{\partial}{\partial t} |\mathcal{E}(z, t)|^2 \Delta z. \tag{2.70}$$

Assuming a Gaussian temporal shape of the pulse $\mathcal{E} \sim \exp\left(-t^2/t_p^2\right)$, it follows that the SPM induced change of the instantaneous frequency satisfies

$$\Delta\omega(t) \sim t e^{-t^2/t_p^2}. \tag{2.71}$$

This reveals that action of SPM on the leading edge ($t < 0$) of the pulse produces a spectral redshift, while a blueshift is produced in the trailing edge ($t > 0$). In summary, SPM generates new spectral content, leading to a broadening of the frequency spectrum of the pulse [20, 41]. Under the above approximations, SPM only affects the temporal phase, leaving the temporal profile $|\mathcal{E}(z, t)|$ unchanged. This ceases to be true when dispersion can no longer be neglected. Assuming, for simplicity, that the pulse is subject to GVD only, it can be shown that the combined action of normal GVD ($\beta_2 > 0$) and SPM leads to the phenomenon of optical wavebreaking [41]. This becomes noticeable as a steepening both of the leading and trailing edges of the pulse, which in turn yields a strong impact of GVD on the steepened pulse edges. The latter causes rapid oscillations of the pulse envelope in leading and trailing edge. The formation of pronounced spectral sidelobes are the frequency domain analog of optical wavebreaking.

If the initial pulse is symmetric in time, the aforementioned interplay of GVD and SPM preserves this symmetry as the pulse propagates along z. However, the situation dramatically changes when the pulse duration approaches the order of the optical cycle. In this case, the operator T in Eq. (2.55) becomes essential for a physically reasonable description, and the dynamical equation describing SPM in the few-cycle regime reads

$$\partial_z \mathcal{E} = -i\frac{\beta_2}{2}\frac{\partial^2}{\partial t^2}\mathcal{E} + i\frac{\omega_0}{c}n_2 T|\mathcal{E}|^2\mathcal{E}. \tag{2.72}$$

Here, the operator T can be regarded to account for an intensity dependence of the group velocity. For positive n_2, it takes into account that more intense parts of the pulse propagate slower and are delayed with respect to the less intense parts. This behavior causes a steepening of the trailing edge of the pulse, while the leading edge is unaffected by steepening effects. This characteristically asymmetric effect is known as self-steepening [48]. Depart of the characteristically asymmetric temporal pulse profile, SPM in the few-cycle domain also leads to a strong asymmetry in the spectrum. In fact, the generation of new frequency components by SPM is most pronounced when the envelope exhibits a strong temporal gradient. Therefore, the generation of blue spectral content in the steepened trailing edge of the pulse is strongly enhanced compared to SPM based generation of red frequencies in the slowly rising leading edge of the pulse. It follows that few-cycle pulses subject to SPM typically exhibit a strongly asymmetric spectrum, with a pronounced blueshifted spectral tail. In addition, as the blue spectral components are dominantly generated in the vicinity of the self-steepened trailing edge of the pulse, they are strongly localized in the temporal domain, leading to a nearly flat spectral phase in the blueshifted spectral range.

2.5.3 Self-Focusing

Besides the modulation of the temporal phase leading to spectral broadening and optical wavebreaking, the IDRI can cause a modulation of the spatial phase of the pulse. Assuming a monochromatic cw-beam propagating in a medium with vanishing plasma response, one finds that in this regime, setting $\mathcal{E} = \mathcal{E}(x, y, z)$, $T = 1$ and $\mathcal{D} = W = \rho = 0$ leads to the simplified Eq. [5]

$$\partial_z \mathcal{E} = \frac{i}{2k_0}\Delta_\perp \mathcal{E} + i\frac{\omega_0}{c}n_2|\mathcal{E}|^2\mathcal{E} \tag{2.73}$$

This is the Nonlinear Schrödinger equation in two transverse spatial dimensions (x, y) and one dimension corresponding to the propagation direction z. It corresponds to the paraxial approximation of the Helmholtz equation, augmented by a contribution due to the IDRI $n = n_0 + n_2 I$. The nonlinear part of the refractive index gives rise to a self-induced modulation of the spatial phase,

$$\phi(x, y, z) \rightarrow \phi(x, y, z) + \frac{\omega_0}{c} n_2 I(x, y, z) \Delta z. \tag{2.74}$$

For a Gaussian beam and positive n_2, the self-induced spatial phase exhibits negative curvature and mimics the action of a focusing lens. This may lead to persistent self-focusing of the beam until the intensity blows up, as was first observed in Ref. [49]. The nonlinear Schrödinger equation can be analyzed by several approximate approaches, considering for instance the propagation of rays in a self-induced refractive index profile, or, alternatively, the method of moments [50]. All these approaches predict that a Gaussian beam in a medium with positive n_2 will tend to self-focus until the amplitude blows up at a finite distance z_{cr}, and the solution diverges, given that its optical power $\sim \int |\mathcal{E}(x, y, z)|^2 dx dy$ exceeds a certain critical value P_{cr}. The numerical value of this critical power differs slightly between the various approximative approaches. An analytical treatment of Eq. (2.74) reveals that the critical power for self-focusing is given by [50–52]

$$P_{cr} = \frac{11.69\lambda^2}{8\pi^2 n_0 n_2}. \tag{2.75}$$

Here, P_{cr} is the optical power a of specific transverse profile, the so-called Townes mode [52], which provides a family of stationary solutions to Eq. (2.73). A more detailed discussion of the mathematical background of self-focusing described by the Nonlinear Schrödinger equation can be found in Appendix A. It can be shown that an arbitrarily shaped beam will not collapse if its optical power P satisfies $P < P_{cr}$ [53]. Therefore, $P > P_{cr}$ provides a necessary, but not sufficient condition for collapse to occur. For Gaussian beams, collapse does not occur unless $P > P_{cr}^G \approx 1.02 P_{cr}$ [50]. Note that a frequently used approximation for the critical power is given by $P_{cr} = \lambda_0^2/(2\pi n_0 n_2)$ which is used throughout the thesis. This approximate value can, as will be shown in Sect. (3.1), be derived using a variational approach (see Appendix A). This variational approach is closely related to the aforementioned method of moments. Also, it has to be emphasized that the Townes profile provides a stationary, yet unstable solution to (2.75)[53–55]. This is due to the fact that any arbitrary small perturbation will either cause collapse or decay of the Townes mode. In this aspect, the Townes solution differs from the familiar fiber solitons [41] in the anomalous dispersion regime $\beta_2 < 0$, which are governed by a $(1+1)$-dimensional NLSE analogous to (2.74) and are unconditionally stable. A semiempirical formula for the finite propagation distance z_{cr} at which a Gaussian beam with beam waist w_0 at $z = 0$ collapses was provided by Marburger [56], pursuant to

$$z_{cr} = \frac{0.376 z_0}{\sqrt{\left(\sqrt{\frac{P_{in}}{P_{cr}}} - 0.852\right)^2 - 0.0219 + \frac{z_0}{f}}}, \tag{2.76}$$

where $z_0 = \pi n_0 w_0^2/\lambda_0$ is the Rayleigh range of a collimated Gaussian beam, $f = R/2$ the focal point, and R is the radius of curvature of the beam wavefront.

A detailed analysis of this formula shows that z_{cr} may be smaller than f such that the beam collapses at a position prior to the position of the linear focus. Thus, in order to distinguish it from the geometrical focus of linear theory, z_{cr} is often termed nonlinear focus. Of course, beam collapse does not occur in physical systems, as counteracting effects come into play as the beam intensity blows up. For example, it has been shown that nonparaxiality and vectorial effects can stop the collapse of the beam [16, 57]. However, these effects become relevant at intensities far beyond the photoionization threshold. While photoionization is currently supposed to be the prevalent mechanism to stop beam collapse, as discussed below, the discussion of Chap. 4 reveals the possible role played by higher-order Kerr nonlinearities. Besides, also GVD turns out to be able to saturate the self-focusing collapse, this will be discussed below, and, in more detail, in Sect. 3.3.

2.5.4 Modulation Instabilities

The Townes mode is a radially symmetric solution to (2.74). However, besides the previously discussed self-focusing instability, solutions of (2.74) suffer from the so-called azimuthal modulation instability, which is able to break the radial symmetry of a given solution. To be more precise, an infinitesimally small, radially **asymmetric** perturbation to the radially **symmetric** initial field will, under certain conditions, be exponentially amplified, leading to a spatial break-up and loss of radial symmetry of the initial solution. Theoretically, this was first observed by Bespalov and Talanov [58] by means of a plane wave propagating in the z direction, being perturbed by a small amplitude wave with nonvanishing transverse wave vector $\vec{k}_\perp = (k_x, k_y)$. In the context of femtosecond filamentation, this phenomenon is known as multifilamentation and is expected to be observed when the input power of the pulsed laser beam strongly exceeds the critical power P_{cr} for self-focusing. Nevertheless, for beam powers up to five critical powers, it has been demonstrated that the onset of multifilamentation can be circumvented by means of suitably aperturing the input beam [45].

In addition, modulation instability is the cause of another phenomenon occurring in the context of femtosecond filamentation, namely the generation of hyperbolic shock-waves, X-waves and conical emission [59]. The latter instability occurs due to the interplay of self-focusing and normal GVD. In fact, it can be shown that GVD with $\beta_2 > 0$ is able to counteract the optical collapse induced by the Kerr nonlinearity [60]. Considering the evolution of long input pulses such that the SVEA ($T = 1$) can applied, and further neglecting plasma response and higher order dispersion β_k for $k > 2$, the envelope Eq. (2.55) reduces to

$$\partial_z \mathcal{E} = \frac{i}{2k_0} \Delta_\perp \mathcal{E} - i \frac{\beta_2}{2} \frac{\partial^2}{\partial t^2} \mathcal{E} + i \frac{\omega_0}{c} n_2 |\mathcal{E}|^2 \mathcal{E}. \tag{2.77}$$

For $\beta_2 > 0$, this is a Nonlinear Schrödinger Equation in $(2+1)$-dimensions with a hyperbolic wave operator $\alpha\Delta_\perp - \gamma\frac{\partial^2}{\partial t^2}$ with $\alpha, \gamma > 0$. The latter equation admits identical stationary solution as Eq. (2.74), namely a monochromatic beam with a transverse beam profile given by the Townes mode. However, the detailed analysis of [61] reveals that a small perturbation to the stationary solution may acquire exponential gain, leading to the formation of so-called X-waves. The latter have recently been related to the phenomenon of conical emission frequently observed in filamentation [59].

2.5.5 Space-Time Focusing

Space-time focusing is not a phenomenon restricted to filamentary propagation. Rather, it describes the impact of linear diffraction on ultrashort, ultra-broadband optical pulses. Diffraction, i.e. the spreading of a laser beam in the transverse plane, is governed by the dispersion relation

$$n^2(\omega)\frac{\omega^2}{c^2} = k_x^2 + k_y^2 + k_z^2. \tag{2.78}$$

In the case of paraxial propagation into the positive z-direction, the latter relation may be approximated using $\sqrt{1+x} \approx 1 + x/2$, which yields

$$k_z = k(\omega)\left(1 - \frac{k_\perp^2}{2k^2(\omega)}\right). \tag{2.79}$$

However, it was noticed in Ref. [62] that the SVEA, which corresponds to the replacement $k(\omega) \rightarrow k_0$ in the latter equation, fails to describe spatiotemporal coupling effects which occur for ultrashort, diffracting laser pulses. Indeed, it can be deduced from the dispersion relation Eq. (2.78) that blue spectral components of a pulse diffract slower than their red spectral counterparts, which eventually gives rise to a narrowing of the on-axis spectrum of the beam. This corresponds to a pulse broadening in the temporal domain. Therefore, it became necessary to include a correction term to the SVEA model to account for the spatiotemporal coupling effect governing the diffraction of ultrashort pulses. In the NEE model (2.55), this correction is provided by the augmented diffraction term $\sim T^{-1}\Delta_\perp$. In contrast to the NEE model, the factorization procedure leading to the FME naturally respects the dispersion relation (2.79).

2.5.6 *Intensity Clamping and the Dynamic Spatial Replenishment Model*

The intensity clamping model [63] of femtosecond filamentation assumes that the onset of photoionization and the subsequent defocusing effect due to free carriers is the dominant mechanism to arrest the beam collapse triggered by Kerr self-focusing. Using the expression for the IDRI $\Delta n = n_2 I$ and the corresponding expression for the plasma-induced refractive index change according to Eq. (2.43), the nonlinear change of the refractive index is given by

$$\Delta n = n_2 I - \frac{\rho}{n_0^2 2 \rho_c} \tag{2.80}$$

For a self-focusing medium $n_2 > 0$, the plasma contribution to the refractive index has a sign opposite to that of the Kerr nonlinearity. Therefore, the plasma-induced refractive index profile acts like a defocusing lens, as the density of free carriers generally increases towards the optical axis. The intensity clamping model provides an estimate for the maximal intensity achieved within femtosecond filaments. At the clamping intensity, defined as the solution of

$$\Delta n \equiv n_2 I - \frac{\rho}{n_0^2 2 \rho_c} = 0, \tag{2.81}$$

the plasma induced refractive index change is sufficient to generate a defocusing effect that balances Kerr induced self-focusing. Using (2.56) and neglecting plasma losses, the rate equation for ρ can be integrated to yield

$$\rho = \rho_0 \left(1 - \exp\left[- \int_{-\infty}^{t} w[I(t')]dt' \right] \right). \tag{2.82}$$

However, in order to obtain an estimate of the clamping intensity, it suffices to consider the peak plasma density and peak pulse intensities, $\rho_{\max}$ and $I_{\max}$. Using $\exp(-x) \approx 1 - x$ for not too high plasma densities, one finds that $\rho_{\max} = t_p \rho_0 w(I_{\max})$, leading to the following equation for the clamping intensity [8]

$$I_{\max} = \frac{t_p \rho_0 w(I_{\max})}{2 \rho_c n_0 n_2}. \tag{2.83}$$

It has been shown that this expression generally provides a good estimate for the intensity in the nonlinear focus [13]. However, intensity clamping does not imply that long-range filamentary propagation is a stationary process, with constant intensity $I_{\max}$ along the entire longitudinal extension of the filament. Rather, filamentation is a highly dynamical process, and a detailed examination of a femtosecond filament reveals that it consists of recurrent nonlinear foci along its longitudinal axis. Besides,

it has been demonstrated in Ref. [64] that the different temporal sections of the pulse
are subject to repeated focusing and refocusing cycles such that the pulse fluence, i.e.
the time integrated intensity given in units of J/m^2, appears stationary and, according
to [64]: "[...]produce[s] the illusion of long-distance propagation of one self-guided
pulse". This model was termed Dynamic spatial replenishment by the authors of [64].

2.5.7 Pulse Self-Compression

Pulse self-compression probably is the most intriguing phenomenon observed in
femtosecond filamentation. Unless traditional schemes for laser pulse compression,
it enables the generation of ultrashort pulses consisting only of a few cycles of the
optical carrier field without the need for any dispersive compression techniques. The
following Chap. 3 will discuss filamentary self-compression in detail, revealing both
theoretical and experimental aspects of the topic.

References

1. J.C. Maxwell, On physical lines of force. Philos Mag. **21**, 161(1861)
2. A. Ferrando, M. Zacarés, P. Fernández de Córdoba, D. Binosi, A. Montero, Forward-backward
   equations for nonlinear propagation in axially invariant optical systems. Phys. Rev. E **71**,
   016601 (2005)
3. P. Kinsler, Optical pulse propagation with minimal approximations. Phys. Rev. A **81**, 013819
   (2010)
4. A.V. Husakou, J. Herrmann, Supercontinuum Generation of Higher-Order Solitons by Fission
   in Photonic Crystal Fibers. Phys. Rev. Lett. **87**, 203901 (2001)
5. V.E. Zakharov, A.B. Shabat, Exact theory of twodimensional selffocusing and one-dimensional
   self-modulation of waves in nonlinear media. Sov. Phys. JETP **101**, 62 (1972)
6. T. Brabec, F. Krausz, Nonlinear optical pulse propagation in the single-cycle regime. Phys.
   Rev. Lett. **78**, 3282 (1997)
7. S. Skupin, G. Stibenz, L. Berge, F. Lederer, T. Sokollik, M. Schnürer, N. Zhavoronkov,
   G. Steinmeyer, Self-compression by femtosecond pulse filamentation: Experiments versus
   numerical simulations. Phys. Rev. E **74**, 056604 (2006)
8. L. Berge, S. Skupin, R. Nuter, J. Kasparian, J.P. Wolf, Ultrashort filaments of light in weakly
   ionized, optically transparent media. Rep. Prog. Phys. **70**, 1633 (2007)
9. S. Augst, D. Strickland, D.D. Meyerhofer, S.L. Chin, J. Eberly, Tunneling ionization of noble
   gases in a high-intensity laser field. Phys. Rev. Lett. **63**, 2212 (1989)
10. V.S. Popov, Tunnel and multiphoton ionization of atoms and ions in a strong laser field. Phys.
    Usp. **47**, 855 (2004)
11. A. Braun, G. Korn, X. Liu, D. Du, J. Squier, G. Mourou, Self-channeling of high-peak-power
    femtosecond laser pulses in air. Opt. Lett. **20**, 73 (1995)
12. A. Couairon, A. Mysyrowicz, Femtosecond filamentation in transparent media. Phys. Rep.
    441, 47 (2007)
13. S.L. Chin, Y. Chen, O. Kosareva, V.P. Kandidov, F. Théberge, What is a filament? Laser Phys.
    18, 962 (2008)

14. L.D. Landau, E.M. Lifschitz. Lehrbuch der Theoretischen Physik, Bd. 8, Elektrodynamik der Kontinua, (Harri Deutsch, Berlin, 1991)
15. V.M. Agranovich, Y.N. Gartstein, Spatial dispersion and negative refraction of light. Phys. Usp **49**, 1029 (2006)
16. G. Fibich, B. Ilan, Deterministic vectorial effects lead to multiple filamentation. Opt. Lett. **26**, 840 (2001)
17. M. Kolesik, J.V. Moloney, M. Mlejnek, Unidirectional optical pulse propagation equation. Phys. Rev. Lett. **89**, 283902 (2002)
18. S. Amiranashvili, A.G. Vladimirov, U. Bandelow, A model equation for ultrashort optical pulses. Eur. Phys. J. D **58**, 219 (2010)
19. S. Amiranashvili, A. Demircan, Hamiltonian structure of propagation equations for ultrashort optical pulses. Phys. Rev. A **82**, 013812 (2010)
20. R.W. Boyd, *Nonlinear Optics*, (Academic Press, Orlando, 2008)
21. S. Skupin, O. Bang, D. Edmundson, W. Krolikowski, Stability of two-dimensional spatial solitons in nonlocal nonlinear media. Phys. Rev. E **73**, 066603 (2006)
22. D.C. Hutchings, M. Sheik-Bahae, D.J. Hagan, E.W. van Stryland, Kramers-Kronig relations in nonlinear optics. Opt. Quant. Electron. **24**, 1 (1992)
23. W.R.J.C. Diels, *Ultrashort Laser Pulse Phenomena: Fundamentals, Techniques, and Applications on a Femtosecond Time Scale*, (Academic Press, Burlington, 2006)
24. C. Brée, A. Demircan, G. Steinmeyer, Method for computing the nonlinear refractive index via Keldysh theory. IEEE J. Quantum Electron. **4**, 433 (2010)
25. P.P. Ho, R.R. Alfano, Optical Kerr effect in liquids. Phys. Rev. A **20**, 2170 (1979)
26. M. Melnichuk, L.T. Wood, Direct Kerr electro-optic effect in noncentrosymmetric materials. Phys. Rev. A **82**, 013821 (2010)
27. P. Bejot, J. Kasparian, S. Henin, V. Loriot, T. Vieillard, E. Hertz, O. Faucher, B. Lavorel, J.-P. Wolf, Higher-order Kerr terms allow ionization-free filamentation in gases. Phys. Rev. Lett. **104**, 103903 (2010)
28. V. Loriot, E. Hertz, O. Faucher, B. Lavorel, Measurement of high order Kerr refractive index of major air components: erratum. Opt. Express **18**, 3011 (2010)
29. V. Loriot, E. Hertz, O. Faucher, B. Lavorel, Measurement of high order Kerr refractive index of major air components. Opt. Express **17**, 13429 (2009)
30. P. Drude, Zur Elektronentheorie der Metalle. Annalen der Physik, **306**, 566 (1900). ISSN 1521-3889
31. P. Drude, Zur Elektronentheorie der Metalle; II. Teil. Galvanomagnetische und thermomagnetische Effecte. Annalen der Physik, **308**, 369 (1900). ISSN 1521-3889
32. A.M. Perelomov, V.S. Popov, M.V. Terent'ev, Ionization of atoms in an alternating electric field. Sov. Phys. JETP **23**, 924 (1966)
33. M.V. Ammosov, N.B. Delone, V.P. Krainov, Tunnel ionization of complex atoms and of atomic ions in an alternating electromagnetic field. Sov. Phys. JETP **64**, 1191 (1986)
34. L.V. Keldysh, Ionization in the field of a strong electromagnetic wave. Sov. Phys. JETP **20**, 1307 (1965)
35. S.V. Poprzuhenko, V.D. Mur, V.S. Popov, D. Bauer, Strong field ionization rate for arbitrary laser frequencies. Phys. Rev. Lett. **101**, 193003 (2008)
36. Y.V. Vanne, A. Saenz, Exact Keldysh theory of strong-field ionization: residue method versus saddle-point approximation. Phys. Rev. A **75**, 033403 (2007)
37. I. Koprinkov, Ionization variation of the group velocity dispersion by high-intensity optical pulses. Appl. Phys. B **79**, 359 (2004). ISSN 0946-2171.10.1007/s00340-004-1553-z
38. Y.H. Chen, S. Varma, T.M. Antonsen, H.M. Milchberg, Direct measurement of the electron density of extended femtosecond laser pulse-induced filaments. Phys. Rev. Lett. **105**, 215005 (2010)
39. J. Kasparian, P. Béjot, J.-P. Wolf, Arbitrary-order nonlinear contribution to self-steepening. Opt. Lett. **35**, 2795 (2010)
40. W. Ettoumi, P. Béjot, Y. Petit, V. Loriot, E. Hertz, O. Faucher, B. Lavorel, J. Kasparian, J.-P. Wolf, Spectral dependence of purely-Kerr-driven filamentation in air and argon. Phys. Rev. A **82**, 033826 (2010)

41. G.P. Agrawal, *Nonlinear Fiber Optics*, 3rd edn. (Academic Press, London, 2001)
42. A. Dalgarno, A.E. Kingston, The refractive indices and Verdet constants of the intert gases. Proc. Roy. Soc. A **259**, 424 (1960)
43. S. Amiranashvili, U. Bandelow, A. Mielke, Padé approximant for refractive index and nonlocal envelope equations. Opt. Commun. **283**, 480 (2010)
44. L. Bergé, S. Skupin, Few-cycle light bullets created by femtosecond filaments. Phys. Rev. Lett. **100**, 113902 (2008)
45. G. Stibenz, N. Zhavoronkov, G. Steinmeyer, Self-compression of milijoule pulses to 7.8 fs duration in a white-light filament. Opt. Lett. **31**, 274 (2006)
46. J.-F. Daigle, O. Kosareva, N. Panov, M. Bégin, F. Lessard, C. Marceau, Y. Kamali, G. Roy, V. Kandidov, S.L. Chin, A simple method to significantly increase filaments' length and ionization density. Appl. Phys. B **94**, 249 (2009)
47. R.R. Alfano, S.L. Shapiro, Observation of self-phase modulation and small-scale filaments in crystals and glasses. Phys. Rev. Lett. **24**, 592 (1970)
48. F. DeMartini, C.H. Townes, T.K. Gustafson, P.L. Kelley, Self-steepening of light pulses. Phys. Rev. **164**, 312 (1967)
49. G.A. Askar'yan, Effects of the gradient of strong electromagnetic beam on electrons and atoms. Sov. Phys. JETP **15**, 1088 (1962)
50. L. Bergé, Wave collapse in physics: principles and applications to light and plasma waves. Phys. Rep. **303**, 259 (1998)
51. C. Sulem, P.-L. Sulem, *The Nonlinear Schrödinger Equation: Self-Focusing and Wave Collapse*, Applied Mathematical Sciences(Springer-Verlag, New York, 1999)
52. R.Y. Chiao, E. Garmire, C.H. Townes, Self-trapping of optical beams. Phys. Rev. Lett. **13**, 479 (1964)
53. M.I. Weinstein, Nonlinear Schrödinger equations and sharp interpolation estimates. Commun. Math. Phys. **87**, 567 (1983)
54. J.J. Rasmussen, K. Rypdal, Blow-up in nonlinear Schrödinger equations-I: a general review. Phys. Scr. **33**, 481 (1986)
55. K. Rypdal, J.J. Rasmussen, Stability of solitary structures in the nonlinear Schrödinger equation. Phys. Scr. **40**, 192 (1989)
56. J. Marburger, Self-focusing: theory, Prog. Quant. Electron. **4**, 35 (1975). ISSN 0079-6727
57. G. Fibich, Small beam nonparaxiality arrests self-focusing of optical beams. Phys. Rev. Lett. **76**, 4356 (1996)
58. V.I. Bespalov, V.I. Talanov, Filamentary structure of light beams in nonlinear liquids. J. Exp. Theor. Phys. **11**, 471 (1966)
59. D. Faccio, M.A. Porras, A. Dubietis, F. Bragheri, A. Couairon, P.D. Trapani, Conical emission, pulse splitting, and x-wave parametric amplification in nonlinear dynamics of ultrashort light pulses. Phys. Rev. Lett. **96**, 193901 (2006)
60. L. Bergé, J.J. Rasmussen, Multisplitting and collapse of self-focusing anisotropic beams in normal/anomalous dispersive media. Phys. Plasmas **3**, 824 (1996)
61. M.A. Porras, A. Parola, D. Faccio, A. Couairon, P.D. Trapani. Light-filament dynamics and the spatiotemporal instability of the Townes profile. Phys. Rev. A **76**, 011803(R) (2007)
62. J.E. Rothenberg, Space-time focusing: breakdown of the slowly varying envelope approximation in the self-focusing of femtosecond pulses. Opt. Lett. **17**, 1340 (1992)
63. A. Becker, N. Aközbek, K. Vijayalakshmi, E. Oral, C. Bowden, S. Chin. Intensity clamping and re-focusing of intense femtosecond laser pulses in nitrogen molecular gas. Appl. Phys. B **73**, 287 (2001). ISSN 0946-2171.10.1007/s003400100637
64. M. Mlejnek, E.M. Wright, J.V. Moloney, Dynamic spatial replenishment of femtosecond pulses propagating in air. Opt. Lett. **23**, 382 (1998)

Chapter 3
Pulse Self-Compression in Femtosecond Filaments

In the current chapter, various theoretical and experimental aspects of pulse self-compression in femtosecond filaments are discussed. The main message of the theoretical part is that filamentary self-compression relies on fundamentally different mechanisms than traditional pulse compression schemes in single-mode-, microstructure-, photonic crystal- and gas filled hollow fibers [1–4]. As a necessary prerequisite for pulse compression, traditional schemes rely on spectral broadening due to self-phase modulation in a fiber with nonvanishing Kerr nonlinearity. However, both linear dispersion and the Kerr nonlinearity introduce a positive chirp on the pulse such that in the frequency domain, the complex envelope of the spectrally broadened pulse is given by

$$\widehat{\mathcal{E}}(\omega) \sim \exp\left(-\frac{\omega^2}{\Delta\omega^2} + i\phi(\omega)\right) \tag{3.1}$$

where a Gaussian spectral distribution with a spectral bandwidth $(1/e)$ $\Delta\omega$ is assumed. The FWHM spectral width $\Delta\omega_{\mathrm{FWHM}}$ is then given by $\Delta\omega_{\mathrm{FWHM}} = \sqrt{2\ln 2}\Delta\omega$. The spectral phase is given by $\phi(\omega)$. Assuming that the medium dispersion is dominated by second-order dispersion β_2, the assumption of a parabolic spectral phase $\phi(\omega) = \alpha\omega^2/2$ is justified. In this case, the group-delay dispersion $D_2 \equiv d^2\phi/d\omega^2 = \alpha$ is frequency-independent and given by the parameter α. A Fourier transform of Eq. (3.1) yields the complex envelope in the time domain,

$$\mathcal{E}(t) = \exp\left(-\frac{t^2}{t_p^2(\alpha)} - iC\frac{t^2}{t_p^2(\alpha)}\right). \tag{3.2}$$

The parameter $t_p(\alpha)$ is related to the FWHM pulse duration according to $t_p(\alpha) = \sqrt{2\ln 2}t_{\mathrm{FWHM}}$, while $C = \alpha\Delta\omega^2/2$ is the chirp factor. The pulse duration depends on the group delay dispersion α according to

C. Brée, *Nonlinear Optics in the Filamentation Regime*, Springer Theses,
DOI: 10.1007/978-3-642-30930-4_3, © Springer-Verlag Berlin Heidelberg 2012

$$t_p(\alpha) = t_p(0)\sqrt{1 + \frac{4\alpha^2}{t_p^4(0)}}. \qquad\qquad (3.3)$$

Here, $t_p(0) = 2/\Delta\omega$ is the duration of the corresponding transform-limited pulse, i.e. the pulse obtained by setting $\alpha = 0$ in Eq. (3.1), yielding a flat spectral phase over the whole frequency range. For the given spectral bandwidth $\Delta\omega$, $t_p(0) = 2/\Delta\omega$ is the shortest attainable pulse duration, assuming a Gaussian spectral shape. Since the GDD acquired by the pulse during the process of spectral broadening leads to a significant temporal stretching according to Eq. (3.3), it is the aim of traditional pulse compression techniques to apply suitable dispersion compensation techniques that eliminate the GDD in Eq. (3.1) in order to obtain a short, transform limited pulse. In an experimental setup, this may be achieved by suitably aligning a pair of gratings or chirped mirrors [5–8] which introduce an appropriate group delay between the different spectral components of the pulse so as to produce negative GDD. For an appropriately chosen compressor geometry, this amount of negative GDD precisely cancels the positive GDD α of the spectrally broadened pulse, resulting in the desired transform-limited, ultrashort pulse. The commonly used dispersive pulse compressors are restricted to compensation of positive second-order dispersion (GDD), i.e. a parabolic spectral phase with positive curvature. In practice, however, ultrashort pulses in optical fibers are also subject to higher-order dispersion such that compensating only for GDD can lead to unwanted satellite pulses and even pulse splitting effects, thus limiting the effectiveness of dispersive pulse compression techniques [9]. This can be avoided, at least partially, by compression schemes which allow for a simultaneous compensation of GDD and TOD (third-order dispersion, β_3) [10]. In summary, it is obvious that these traditional compression schemes reshape the pulse by introducing a temporal energy flow: the positively chirped, temporally stretched pulse is dispersed in time, and each frequency component of the pulse approaches a hypothetical observer at a different time. The dispersion compensation then introduces negative GDD on the spectral phase, such that the energy contained in the different frequency components flows back towards the pulse center, leading to a temporal recompression of the pulse.

Pulse compression in a femtosecond filament was first evidenced in Ref. [8]. However, the experimental setup still included a pair of chirped mirrors for dispersion compensation. Pulse self-compression without any dispersion compensation scheme is first described in Ref. [11]. Besides the fact that no external dispersion compensation scheme is necessary in filamentary self-compression, this pulse compression method does not require any guiding structures such as fibers or hollow fibers and is therefore not limited by damage threshold. Instead, interacting nonlinear effects lead to a self-guiding effect in the propagation medium as will be detailed below. Simultaneously, under appropriate initial conditions, the on-axis temporal profile of the pulse will shorten. For the moment, it will be assumed that the relevant mechanisms acting in this pulse self-compression scenario are analogous to those in traditional compression schemes. This requires the existence of an intrinsic source of negative GVD within the propagation medium. Regarding the fact that the noble gases He,

Ne, Ar, Kr, Xe exhibit positive β_2 in the wavelength regions typically used in filamentation, such source of negative GVD is not straightforward to figure out. One possible candidate for a source of negative GVD may be the plasma generated by photoionization. According to the Drude-Theory, the refractive index of a partially ionized medium is given by

$$n(\omega) = \sqrt{n_0^2(\omega) - \frac{\omega_p^2}{\omega}}, \tag{3.4}$$

where n_0 is the refractive index of the neutral medium and $\omega_p = \sqrt{Ne^2 \epsilon_0 m_e}$ is the plasma frequency depending on the number N of free electrons. Using $k(\omega) = n(\omega)\omega/c$ and $\beta_2(\omega) = d^2 k(\omega)/d\omega^2$, an expression accounting for the plasma GVD has been derived in Ref. [12] according to

$$\beta_{2,e} = -1.58 \times 10^{-5} \lambda^3 N \frac{\text{fs}^2}{\text{cm}}. \tag{3.5}$$

The wavelength λ and carrier density N are given in units of cm and cm^{-3}, respectively. Using the typical wavelength of a Ti:sapphire laser, $\lambda = 800\,\text{nm}$ and a fraction of approximately 0.1 % of ionized particles, the latter being the characteristic plasma density achieved in a femtosecond filament, a plasma GVD of $\beta_{2,e} = -0.22\,\text{fs}^2/\text{cm}$ is obtained. In contrast, the GVD of argon at 800 nm amounts to $\beta_{2,0} = 0.2\,\text{fs}^2/\text{cm}$. In the following example, the hypothesis that filamentary self-compression acts completely analogous to traditional compression mechanisms will be tested. To this aim, self-compression of a Gaussian input pulse with an initial duration $t_p = 40\,\text{fs}$ will be analyzed, presupposing nonlinear spectral broadening and dispersive pulse shaping processes as the driving forces behind filamentary self-compression. It is first assumed that the pulse is subject to spectral broadening such that the resulting spectrum supports a transform-limited pulse duration of $t_p(0) = 10\,\text{fs}$, a realistic compression ratio achievable in self-compression experiments [11]. Furthermore, in analogy to fiber based compression schemes, one might expect that spectral broadening during filamentary propagation is accompanied with additional positive GDD accumulated by the spectral phase of the pulse. Assuming, for instance, an acquired GDD of $\approx 194\,\text{fs}^2$, it follows that the initial pulse duration is unchanged. Thus, in order to explain filamentary self-compression in terms of dispersive pulse shaping mechanisms as in traditional compression schemes, it has to be assumed that the positive excess GDD is compensated by some source of negative GDD. In the following, this is assumed to stem from plasma contributions to the refractive index which provide the necessary dispersion compensation to yield a transform limited, self-compressed 10 fs-pulse. However, in order to ensure that the plasma GVD $\beta_{2,e} = -0.2\,\text{fs}^2/\text{cm}$ compensates for the acquired positive chirp of $\approx 200\,\text{fs}^2$, the pulse has to propagate along a distance $\Delta z = 100\,\text{m}$ in a strongly ionized ($N = 2.7 \times 10^{16}\,\text{cm}^{-3}$) self-generated filament, according to $\text{GDD} = \beta_2 \Delta z$. In contrast, the longitudinal extension of the filament in self-compression experiments is typically of the order of only 1 m. Moreover, in these configurations, the length of the strongly ionized plasma channel is of the order of some 10 cm. In consequence, due to the weak GVD both of the gas and the plasma, dispersive pulse shaping can be clearly ruled out as the main mechanism leading to pulse self-compression in femtosecond filaments.This makes it evident that the mechanisms behind filamentary self-compression are unlike those acting in traditional pulse compression methods, and it is the purpose of the subsequent sections to identify these mechanisms.

3.1 The Self-Pinching Mechanism: Self-Compression as a Spatial Effect

Magneto-hydrodynamics (MHD) provides effective mechanisms for increasing the electron density within high-current pulsed discharges. In the plasma channel the self-generated magnetic field may act to radially focus the electron fluence to near-thermonuclear current densities, with the z-pinch [13, 14] being one of the most prominent examples. In contrast, according to the preceding discussion, laser pulse compression [1, 4, 5] has traditionally pursued energy concentration along the temporal coordinate rather than radial contraction. In the following it will be shown that inside a light filament, the combination of only three effects, namely diffraction, Kerr self-focusing, and plasma-induced self-defocusing, holds for a radial contraction mechanism acting on the photon fluence. In analogy to the z-pinching in MHD, this mechanism will be referred to as self-pinching in the following. This phenomenon gives rise to spatio-temporally inhomogeneous configurations of the optical field, implying strong temporal variations of the beam radius [11, 15]. In contrast to previous explanations (see, e.g., [15–17]) of the self-compression in filaments that indicated a complex interplay of some ten effects, in the course of the present section it turns out that only the above-mentioned three spatial effects suffice for self-compression [18]. Propagation of short laser pulses in a filament involves numerous linear and nonlinear optical processes that are typically modeled in the framework of a Nonlinear Schrödinger Equation (NLSE). It is quite remarkable that MHD bears a very similar NLSE for the magnetic field, which may give rise to ionospheric filaments and mechanisms analogous to self-pinching [19]. As all these scenarios exhibit a complex interplay of linear and nonlinear processes it is generally difficult to isolate the primary processes leading to the observed phenomena. For the optical case, however, one can compute characteristic lengths of the participating processes [20] to sort out group-velocity dispersion, absorption, Kerr-type self-phase modulation and self-steepening, leaving mainly plasma effects and transverse self-focusing and -defocusing as suspected drivers behind the experimentally observed self-compression. Such analysis, in particular neglection of dispersion, is indicative of vanishing energy flow along the temporal axis of a pulse in the filament. This essentially leaves particle densities and respective fluences as key parameters, similar to the situation in MHD. The discussion will therefore be restricted to analyzing radial energy flow, for which an extended NLSE in cylindrical coordinates (r, t), assuming cylindrical symmetry, is used [21]. This extended NLSE effectively couples the photon density to the electron density ρ. Compared to the full model Eq. (2.55), energy flow along the t-axis and dissipation are neglected. These effects have been proven unimportant in gaseous media at low or atmospheric pressure both, in previous theoretical and experimental studies [11, 15].

Consequently, a reduced model considering only Kerr-type self-focusing and plasma defocusing as main dynamic effects during filament formation in gases turns out sufficient to observe self-compression:

$$\partial_z \mathcal{E} = \frac{i}{2k_0 r} \partial_r r \partial_r \mathcal{E} + \frac{i\omega_0}{c} n_2 |\mathcal{E}|^2 \mathcal{E} - \frac{i\omega_0}{2n_0 c \rho_c} \rho[I, t]\mathcal{E}, \tag{3.6}$$

$$\rho[I, t] = \rho_0 \left(1 - \exp\left(-\int_{-\infty}^{t} dt' w[I(r, z, t')] \right) \right). \tag{3.7}$$

Here, z is the propagation variable, t the retarded time, and ω_0 is the central laser frequency at $\lambda_0 = 2\pi n_0/k_0 = 800$ nm. n_2 is the nonlinear refraction index. Photon densities are described via the complex optical field envelope $\mathcal{E}$, with $I = |\mathcal{E}|^2$. The wavelength-dependent critical plasma density is calculated from the Drude model according to $\rho_c \equiv \omega_0^2 m_e \epsilon_0/q_e^2$, where q_e and m_e are electron charge and mass, respectively, ϵ_0 is the dielectric constant, c the speed of light, and ρ_0 denotes the neutral density. Plasma generation is driven by the ionization rate $w[I]$, which is suitably described by Perelomov-Popov-Terent'ev (PPT) theory [22]. It is emphasized here that for vanishing plasma density $\rho \equiv 0$, the delay t is a free parameter of Eq. (3.6). This means that the envelopes $\mathcal{E}(r, z, t)$ for each instant t evolve independently. For non-vanishing ρ, Eq. (3.7) models a persistent, noninstantaneous nonlinearity which introduces a memory effect to the evolution Eq. (3.6). Physically, this is owing to the fact that any temporal component of the pulse is affected by the plasma generated by all preceding temporal slices of the pulse, as electron-ion recombination can be neglected on a femtosecond time scale.

For the following investigations, data for argon [15] at atmospheric pressure is used. In Ref. [21], the intensity clamping and the resulting temporal pulse profiles in regimes where a Kerr-induced optical collapse is saturated by plasma defocusing have been analyzed using a time-dependent variational approach. Using appropriate approximations, this analysis gave rise to steady-state temporal profiles with soliton-like qualities. It is the aim of this section to go beyond the approximations of [21] in order to calculate a field configuration that represents a stationary temporal intensity profile maintaining a balance between competing nonlinear effects in every temporal point. As in [21], these are derived from a time-dependent variational approach, with the following trial function

$$\mathcal{E}(r, z, t) = \sqrt{\frac{P(t)}{\pi R^2}} \exp\left[-\frac{r^2}{2R^2} + i\frac{k_0 r^2 \partial_z R}{2R} \right]. \tag{3.8}$$

The quadratic phase guarantees preservation of continuity equations through self-similar substitutions and relates the spatial phase curvature of a Gaussian beam to the logarithmic derivative of the pulse radius w.r.t. the z-coordinate. The pulse radius $R \equiv R(z,t)$ depends on both the longitudinal and temporal variables. Due to the neglection of temporal dispersion and self-steepening terms, no energy flow between different temporal sections of the pulse occurs in the simplified model governed by Eq. (3.6). Furthermore, dissipative effects were ruled out. In consequence, the optical power $P(z,t) = 2\pi \int_0^\infty r \, dr |\mathcal{E}(r,z,t)|^2$ is conserved along propagation, i.e., $\partial_z P(z,t) \equiv 0$. For this conservative system, straightforward algebra provides the virial-type identity [23] (see also Appendix A),

$$\partial_z^2 \int_0^\infty r^3 |\mathcal{E}|^2 dr = \frac{2}{k_0^2} \int_0^\infty r |\partial_r \mathcal{E}|^2 dr - \frac{2n_2}{n_0} \int_0^\infty r |\mathcal{E}|^4 dr - \frac{1}{n_0^2 \rho_c} \int_0^\infty |\mathcal{E}|^2 r^2 \partial_r \rho \, dr.$$

$$(3.9)$$

Inserting the trial function (3.8) with $R(z,t) = w(z,t)/\sqrt{2}$ being related to the Gaussian spot size $w(z,t)$, one obtains a dynamical equation governing the evolution of the pulse radius R along z [23]. For the derivation of analytical expressions for the plasma term on the r.h.s. of Eq. (3.9), the PPT ionization rate is approximated by a power law dependence $w[I] = \sigma_{N*} I_0^{N*}$, with parameters $N* = 6.13$ and $\sigma_{N*} = 1.94 \times 10^{-74} \, \mathrm{s^{-1} cm^{2N*} W^{-N*}}$ fitted to the PPT rate for the intensity range of 80 TW/cm^2. Introducing the **on-axis** intensity profile according to $I_0(t) \equiv I(r = 0, z, t) = P(t)/\pi R^2(t)$, a Gaussian power profile $P(t) = P_{\mathrm{in}} \exp(-2t^2/t_\mathrm{p}^2)$ is imposed with duration t_p and peak input power P_{in} as a boundary condition. This results in the following integral equation for steady state solutions,

$$0 = 1 - \frac{P(t)}{P_{\mathrm{cr}}} + \mu P^2(t) \int_{-\infty}^t dt' \frac{I_0^{N*+1}(t')}{P(t')} \frac{1}{\left(I_0(t) + N* I_0(t') \frac{P(t)}{P(t')}\right)^2}, \qquad (3.10)$$

with $P_{\mathrm{cr}} = \lambda_0^2/(2\pi n_0 n_2)$ and $\mu = k_0^2 N* \sigma_{N*} \rho_0 / \pi \rho_c$.

Equation (3.10) is basically a generalization of a Volterra-Urysohn integral equation [24], with a kernel depending not only on $I_0(t')$ but also on $I_0(t)$. Here, Eq. (3.10) is solved without the approximations made in Ref. [21]. Taking into account that the integral term of Eq. (3.10) is strictly positive, it immediately follows that nontrivial solutions only exist on the temporal interval $-t_* < t < t_*$ where $P(t) > P_{\mathrm{cr}}$, with $t_* = (\ln \sqrt{P_{\mathrm{in}}/P_{\mathrm{cr}}})^{1/2} t_\mathrm{p}$. From a physical point of view, Kerr self-focusing can compensate for diffraction only on this interval, enabling the existence of a stationary state. For computing a stationary solution $I_0(t)$ of the integral equation, the method presented in [25] is applied which combines a Chebyshev approximation of the unknown $I_0(t)$ with a Clenshaw–Curtis quadrature formula [26] for the integral term. As the laser beam parameters, a ratio $P_{\mathrm{in}}/P_{\mathrm{cr}} = 2$ and a pulse duration $t_{\mathrm{FWHM}} = \sqrt{2\ln 2} t_\mathrm{p} \approx 100\,\mathrm{fs}$ are chosen, leading to $t_* \approx 50\,\mathrm{fs}$. The spectrum of

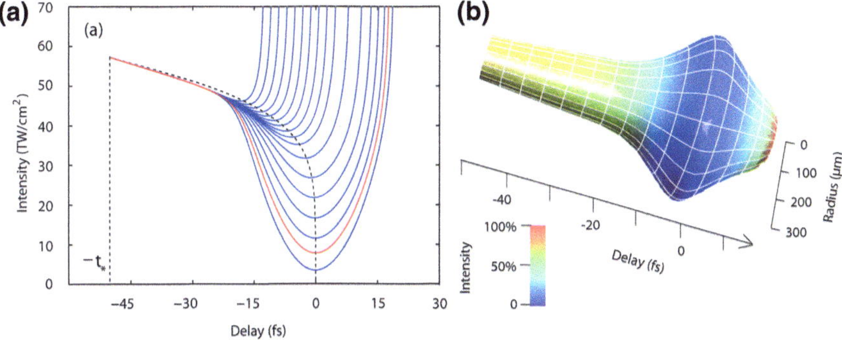

Fig. 3.1 **a** Spectrum of solutions $I_0(t)$ of Eq. (3.10). For the solution $I_0(t)$ marked in *red* [see (**a**)], **b** displays a spatio-temporal representation of the corresponding cylindrically symmetric Gaussian beam, Eq. (3.8), with a time-dependent beam-waist $R(t) = \sqrt{P(t)/\pi I_0(t)}$. $I_0(t)$ solves the steady state condition Eq. (3.10). Color encodes on-axis intensity. Reprinted from C. Brée et al. [18] with permission from Optical Society of America

solutions thus obtained is depicted in Fig. 3.1a. As $1 - P(t)/P_{cr}$ vanishes at the boundaries, there exists a continuum of multiple roots. All solutions show a strongly asymmetric temporal shape, with an intense leading subpulse localized at $t = -t_*$ [21] and a minimum [dashed line in Fig. 3.1a] localized near zero delay, followed by a region of rapid intensity increase, suggesting singular behavior of the solutions. Filamentation is known to proceed from a *dynamical* balance between the Kerr and plasma responses, and a steady-state solution cannot strictly be reached by the physical system. Nevertheless, Eq. (3.10) provides deep insight into the configuration that the pulse profile tends to achieve in the filamentary regime. The rotationally symmetric structure shown in Fig. 3.1b provides a graphical representation of the cylindrically symmetric Gaussian beam Eq. (3.8). The latter exhibits a beam-waist $R(t)$ varying in time, which is encoded by the radius of the solid of revolution at each given instant. This representation allows to identify deviations from a Gaussian beam without any spatio-temporal couplings, i.e., a beam with a time-independent beam-waist. Such deviations are clearly expected for intense optical pulses subject to the interplay of instantaneous Kerr self-focusing and noninstantaneous plasma defocusing.

Indeed, the structure of the emerging solutions (Fig. 3.1b) indeed indicates the formation of two areas of high on-axis intensity being separated by an approximately 20 fs wide defocused zone of strongly reduced intensity. While similar double-peaked on-axis intensities have already been observed in numerical simulations and experiments [11, 27–30] many authors considered a parasitic dispersive break-up in bulk media or optical fibers. Despite its superficial similarity, however, such a break-up cannot be exploited for the compression of isolated femtosecond pulses as will be done below. Interestingly, a comparable dynamical behavior is observed as reported for condensed media, where temporal break-up around zero-delay and the subsequent

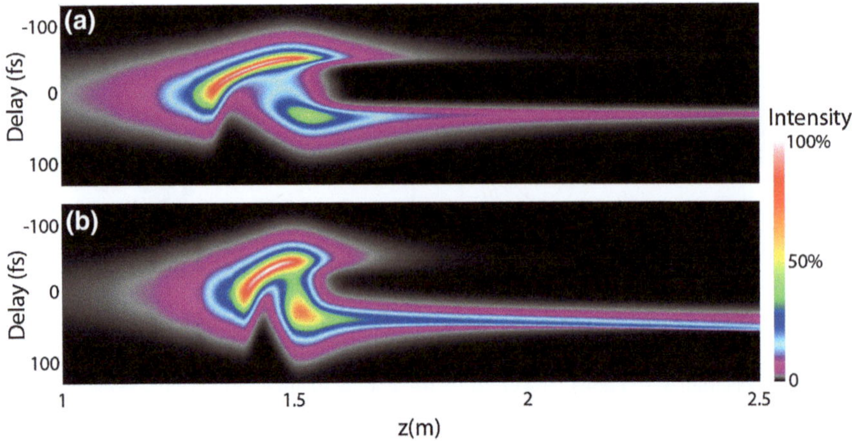

Fig. 3.2 a Evolution of the on-axis temporal intensity profile along z for the reduced numerical model governed by Eq. (3.6). **b** Same for the simulation of the full model equations [15]. Reprinted from C. Brée et al. [18] with permission from Optical Society of America

emergence of nonlinear X-waves occurs. These X-waves were recently proposed to constitute attractors of the filament dynamics [31, 32].

A deeper substantiation of the analytical model is obtained through direct numerical simulations using the reduced radially symmetric evolution model Eq. (3.6) for the envelope of the optical field (see Appendix B for more details on the numerics). The incident field is modeled as a Gaussian in space and time with $w_0 := w(z = 0, t) = 2.5$ mm and identical peak input power and pulse duration as used for the solutions of Eq. (3.10). The field is focused into the medium with an $f = 1.5$ m lens. The result of these simulations can be considered as prototypical for the pulse shaping effect inside filaments. These simulations also demonstrate that spatial effects alone already suffice for filamentary self-compression. As the evolution of the on-axis temporal intensity profile in Fig. 3.2a reveals, filamentary compression always undergoes two distinct phases. Initially, while z approaches the nonlinear focus ($z = 1.4 - 1.5$ m), a dominant leading peak is observed. When the trailing part of the pulse refocuses in the efficiently ionized zone ($\rho_{max} \approx 5 \times 10^{16} \text{cm}^{-3}$) a double-spiked structure emerges. This transient double pulse structure confirms the pulse break-up predicted from the analysis of Eq. (3.10), see Fig. 3.3a, c, and is compatible with the stationary shapes detailed in Fig. 3.1a. Here, analogous to Fig. 3.1, Fig. 3.3c shows a solid of revolution whose radius at each instant t encodes the RMS width of the optical field configuration obtained from the numerics. Subsequently, only one of the emerging peaks survives and experiences further pulse shaping in the filamentary channel.

At $z = 1.7$ m the leading subpulse has already been reduced to a fraction of its original on-axis intensity. This effective attenuation of the leading pulse isolates the trailing pulse that now exhibits a duration $t_{\text{FWHM}} = 27$ fs. The combined split-

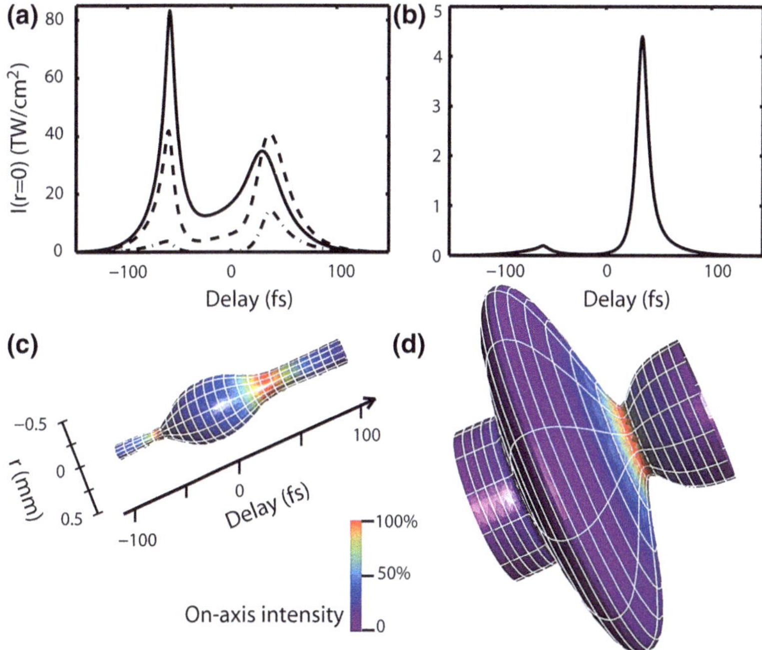

Fig. 3.3 **a** Pulse sequence illustrating the two-stage self-compression mechanism. Shown are the on-axis intensity profiles for $z = 1.5\,$m (*solid line*), $z = 1.55\,$m (*dashed line*) and $z = 1.7\,$m (*dash-dotted line*). **b** Self-compressed few-cycle pulse at $z = 2.5\,$m. **c** Solid of revolution encoding the instantaneous radial RMS width of the optical field exhibiting the double-spiked structure on-axis at $z = 1.55\,$m. **d** Same for the few-cycle pulse at $z = 2.5\,$m. Reprinted from C. Brée et al. [18] with permission from Optical Society of America

up and isolation during the first phase therefore already provides an about fourfold compression of the 100 fs input pulses. In the subsequent weakly ionized zone of the filament ($z > 1.6\,$m), the surviving, trailing subpulse is then subject to additional temporal compression. At $z = 2.5\,$m our simulations indicate pulses as short as $t_{\mathrm{FWHM}} = 13\,$fs (Fig. 3.3b), which agrees favorably with the experimental results in Ref. [11]. In contrast to the plasma-mediated self-compression in the strongly ionized zone, compression in the second zone is solely driven by the Kerr nonlinearity ($\rho < 10^{13}\,$cm^{-3}). With time slices of higher optical powers being able to compensate diffraction by Kerr self-focusing, these portions of the pulse diffract less rapidly than time slices with less optical power. Compared to a linear optical diffraction-ruled scenario, the nonlinear optical effects therefore lead to the formation of a characteristic pinch. This emerging spatial structure is depicted in Fig. 3.3d. This clearly sets the self-pinching mechanism apart from numerous previous reports on seemingly similar dynamics.

For an analytic description of temporal compression during further filamentary propagation, the dynamical equation for the time dependent beam radius derived from Eqs. (3.8) and (3.9) [23] is considered, yet neglecting the plasma term. With the initial conditions $R(z = z_0, t) = R_0$ and $\partial_z R(z = z_0, t) \equiv 0$ the resulting problem is analytically solvable, and one finds

$$R(z, t) = R_0 \sqrt{1 + [(z - z_0)/k_0 R_0^2]^2 (1 - P(t)/P_{cr})}. \tag{3.11}$$

This equation models the evolution of the plasma-free filamentary channel from $z > 1.6\,$m, assuming $P(t) \leq P_{cr}$. The profile $P(t)$ represents the power contained in the filament core region only. For simplicity, it is assumed here that $P(t) = P_{cr} \exp(-2t^2/t_p^2)$, $R_0 = 100\,\mu$m and $t_p = 23\,$fs. This corresponds to the duration of the pulse at $z_0 = 1.7\,$m shown in Fig. 3.3a. Resulting characteristic spatio-temporal shapes are shown in Fig. 3.4. As in Fig. 3.1b, color and radius of the depicted solids of revolution encode instantaneous on-axis intensity $I_0(t)$ and instantaneous beam-waist $R(t)$, respectively. These shapes clearly reveal the presence of self-pinching in this Kerr-dominated stage of propagation and the dominant role it plays for on-axis temporal compression.

The analysis presented so far has completely neglected dispersion, self-steepening and losses. To ensure that dissipation and temporal coupling between time slices have only a modifying effect on the discussed self-compression scenario, full simulations of the filament propagation are performed, including few-cycle corrections and space-time focusing [15]. As shown in Fig. 3.2b, minor parameter adjustment, setting $w_0 = 3.5\,$mm and leaving all other laser beam parameters the same value, suffices to see pulse self-compression within the full model Equation (2.55). Now self-steepening provides a much more effective compression mechanism in the trailing part. However, the comparison of Fig. 3.2a with b also reveals that the dynamical behavior changes only slightly upon inclusion of temporal effects. Clearly, the same two-stage compression mechanism is observed as in the reduced model. One therefore concludes that the pulse break-up dynamics in the efficiently ionized zone is already inherent to the reduced dynamical system governed by Eq. (3.6). Rather than relying on the interplay of self-phase modulation and dispersion as in traditional laser pulse compression, filament self-compression is essentially a spatial effect, conveyed by the interplay of Kerr self-focusing and plasma self-defocusing. This dominance of spatial effects favorably agrees with the spatial replenishment model of Mlejnek et al. [28]. However, the present model indicates previously undiscussed consequences on the temporal pulse structure on axis of the filament, leading to the emergence of the pinch-like structure (Fig. 3.3d) that restricts effective self-compression to the spatial center of the filament [15, 33]. The present analysis confirms the existence of a leading subpulse, in the wake of which the short self-compressed pulse is actually shaped during the first stage of filamentary propagation. This leading structure gives rise to a pronounced temporal asymmetry of self-compressed pulses, which is confirmed in experiments [15].

While qualitatively similar break-up processes have frequently been observed in laser filaments, the preceding analysis identifies the spatially induced temporal break-up as a first step for efficient on-axis compression of an isolated pulse. In the

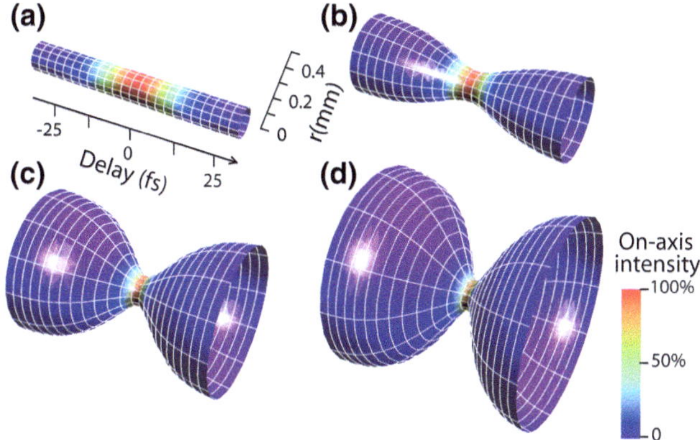

Fig. 3.4 Sequence of pulses illustrating temporal self-compression due to Kerr-induced spatial self-pinching in the variational model corresponding to **a** $z = 1.7$ m, **b** $z = 1.9$ m, **c** $z = 2.1$ m and **d** $z = 2.3$ m. Reprinted from C. Brée et al. [18] with permission from Optical Society of America

current scenario, the leading break-up portion is eventually observed to diffract out and to reduce its intensity, while the trailing pulse can maintain its peak intensity. A subsequent stage, dominated by diffraction and Kerr nonlinearity, serves to further compress the emerging isolated pulse, and may give rise to almost tenfold on-axis pulse compression. The main driver behind this complex scenario is a dynamic interplay between radial effects, namely diffraction, Kerr-type self-focusing, and, exclusively close to the geometric focus, plasma defocusing. The dominance of spatial effects clearly indicates the unavoidability of a pronounced spatio-temporal pinch structure of self-compressed pulses. The frequently observed pedestals in this method are identified as remainders of the suppressed leading pulse from the original split-up. The present analysis also indicates that lower pulse energies <1 mJ requiring more nonlinear gases or higher pressures will see an increased influence of dispersive coupling, which can eventually render pulse self-compression difficult to achieve. Higher energies, however, may not see such limitation, opening a perspective for future improvement of few-cycle pulse self-compression schemes.

3.2 Stationary Solutions Beyond the Variational Approach

The variational approach used in the previous section was successfully used to predict a plasma-induced temporal break-up of the pulse. However, filamentary pulses typically exhibit strong spatio-temporal couplings and, moreover, tend to be reshaped into a system of spatial rings by plasma defocusing and/or conical emission. Therefore, the assumption of a fixed Gaussian radial beam shape used in the variational model governed by Eqs. (3.8), (3.9) requires justification. In the following, this is achieved by comparing the predictions from the variational approach to those extracted from

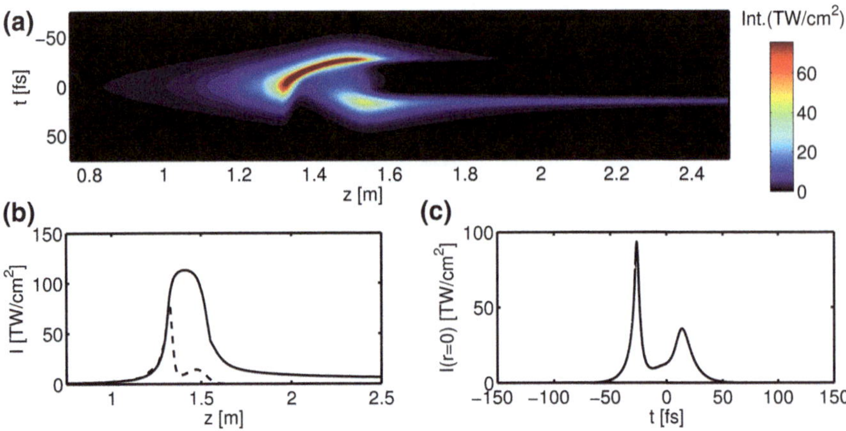

Fig. 3.5 **a** Evolution of the on-axis temporal profile along z. As soon as plasma defocusing has saturated the optical collapse, a characteristic temporal break-up occurs. **b** Evolution of the peak intensity (*solid line*) and the on-axis intensity at zero delay (*dashed line*). **c** On-axis temporal distribution at $z = 1.55$ m exhibiting a typical double hump structure. Reprinted from C. Brée et al. [34]. Copyright © 2010 Astro, Ltd.

calculating stationary solutions directly from the generalized Nonlinear Schrödinger Equation (3.6). In fact, it turns out that the local minimum between the sub-pulses of the characteristic double-hump solutions are again located near the instant where the conserved power profile of the pulse has its maximum, i.e., at zero temporal delay. Thus, the obtained stationary solutions show good qualitative agreement with the solutions calculated from the variational approach [34]. For completeness, these results are then compared with those of direct numerical simulations, where as in the preceding section, only spatial effects are included. This disregards dissipative terms and energy exchange between adjacent temporal slices, thereby ensuring that these effects do not contribute to the observed temporal break-up. The propagation equation of this reduced model is expressed by the NLSE coupled to an evolution equation for the electron density Eqs. (3.6) and (3.7). For the numerical simulations and the analytical discussion, again data for argon [15] at atmospheric pressure is used as medium parameters. As initial conditions for the numerical simulations, a Gaussian spatio-temporal distribution for the photon density is imposed, with a beam waist $w_0 = 2.5$ mm. In contrast to the previous section, where a pulse duration of $t_p = 100$ fs was chosen, here an input pulse duration of $t_p = 38$ fs was chosen in order to adjust to the experimental conditions of [11, 15].

The input energy is $E_{\text{in}} = 1$ mJ, corresponding to a peak input power of $P = 2P_{\text{cr}}$, where $P_{\text{cr}} \approx \lambda^2/2\pi n_2$ is the critical power for Kerr self-focusing. The beam is focused into the medium with an $f = 1.5$ m lens. The evolution of the on-axis intensity depicted in Fig. 3.5a reveals the crucial role of a plasma mediated temporal break-up for an efficient temporal compression induced by local contraction of the spatial beam profile. In fact, the observed compression dynamics is again governed by the

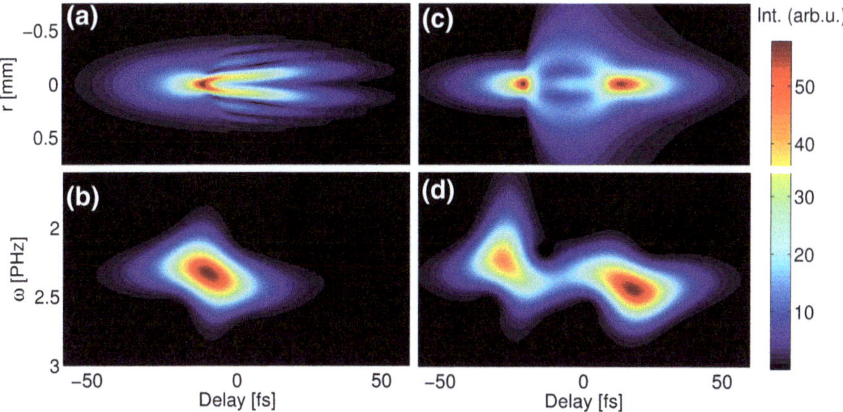

Fig. 3.6 **a** Intensity distribution in the (r, t)-plane as plasma defocusing initiates the pulse break-up at $z = 1.37$ m. **b** Corresponding spectrogram. **c** Intensity distribution of the split pulse at $z = 1.55$ m in the (r, t)-plane. **d** Corresponding spectrogram. Reprinted from C. Brée et al. [34]. Copyright © 2010 Astro, Ltd.

self-pinching mechanism and qualitatively agrees with that of the previous section, cf. Fig. 3.2a. The pulse breaking is initiated when plasma defocusing starts to saturate the optical collapse (see the evolution of the peak intensity in Fig. 3.5b, solid line). Figure 3.6a depicts the intensity distribution in the (t, r) plane at $z = 1.37$ m, clearly revealing defocusing of the trailing part into a system of rings. The corresponding XFROG (cross-correlation frequency-resolved optical gating) spectrogram [15] (cf. Appendix C) of the on-axis intensity profile (Fig. 3.6b), calculated with a 10 fs Gaussian reference pulse, exhibits an inclination due to the generation of red and blue frequencies in the leading and trailing edge of the pulse, respectively. Upon further propagation the rear part of this system of spatial rings merges during a refocusing stage at $z = 1.55$ m, and a blue-shifted trailing subpulse is generated (Fig. 3.6c, d). The on-axis temporal profile of this strongly asymmetric temporal distribution depicted in Fig. 3.5c shows a characteristic double-hump configuration with a local minimum at zero delay. In the simulations, one can track the origin of this minimum to the fact that defocusing prevails at zero delay. Therefore, the energy contained in the spatial rings at zero delay is not transferred back to the optical axis (dashed line in Fig. 3.5b). From the spectrogram representation (Fig. 3.6d) of the split pulse, it becomes obvious that the trailing subpulse is blue-shifted with respect to the leading pulse. This spectro-temporal split is a characteristic feature of filamentary propagation [27, 28, 35], and it is important to note that this split is already fully explicable within the framework of the reduced model equation that incorporates only spatial effects. At first sight, the emergence of the central minimum and the resulting double-hump temporal shapes may appear as a somewhat arbitrary intermediate stage in the pulse shaping process. For a clarification of the role of these characteristic pulse shapes which tend to appear when plasma defocusing saturates Kerr-driven optical collapse, we search for field configurations representing stationary states. These stationary spatio-temporal field distributions maintain a balance between competing nonlinear effects in each time instant. The following analysis circumvents the limiting constraint of a fixed Gaussian radial shape, which has to be imposed in the time-dependent variational approach carried out in the previous section. To further facilitate the calculation of stationary states to the evolution Eq. (3.6), for the ionization rate the multiphoton description $w[I] = \sigma_{N_*}|\mathcal{E}|^{2N^*}$ is used. Here, σ_{N_*} is the cross-section for N^*-photon ionization [36]. As the relevant intensity level in argon filaments is well above the validity of a perturbative multiphoton description, the numerical value of $\sigma_{N^*} = 1.94 \times 10^{-74}\,\mathrm{cm}^{2N^*}\mathrm{W}^{-N^*}$ and $N^* = 6.13$ are determined by a local fit to the ionization rate given by PPT theory. As the model utilized here completely neglects dispersion, the time variable can be regarded as a parameter. Hence the most general ansatz for the stationary state reads as

$$\mathcal{E} = R_{\mu(t)}(r, t)\exp\left(i\mu(t)z\right), \tag{3.12}$$

where an explicit time-dependence of the propagation constant μ is allowed. Substituting this into the dynamical Eq. (3.6) yields the following nonlinear differential equation (cf. Appendix A),

$$0 = \frac{1}{2k_0 r}\partial_r r \partial_r R_{\mu(t)} + \frac{\omega_0}{c} n_2 R^3_{\mu(t)} - \frac{1}{2n_0\rho_c}\frac{\omega_0}{c}\rho R_\mu - \mu(t)R_{\mu(t)}. \tag{3.13}$$

Any solution $R_{\mu(t)}$ of this equation depends on the specific choice of $\mu(t)$, as does the conserved optical power $P = 2\pi \int_0^\infty r\,dr\,R^2_{\mu(t)}$, except for vanishing plasma density $\rho \equiv 0$. For the latter plasma-free case, the solution of (3.13) corresponds to the spatial Townes soliton [37] with an optical power $P'_{\mathrm{cr}} \approx 11.69\lambda^2/(8\pi^2 n_2)$, independent of the chosen value of μ. Note that the optical power of the Townes soliton slightly differs from the usual definition of the critical power $P_{\mathrm{cr}} = \lambda^2/(2\pi n_2)$, with $P_{\mathrm{cr}}/P'_{\mathrm{cr}} \approx 1.075$. In the presence of plasma, the general solution of (3.13) requires introduction of a cut-off time $-t_*$, imposing $P(t) < P'_{\mathrm{cr}}$ for $t < -t_*$ similar to the variational analysis in the previous section. With this constraint, Kerr self-focusing cannot compensate for linear diffraction at $t < -t_*$, and neither can a nontrivial stationary state exist. The solution at $t = -t_*$ itself radially coincides with the Townes soliton, as we assume $\rho \equiv 0$ at this very instant. Imposing a Gaussian power profile $P = P_{\mathrm{in}}\exp\left(-2t^2/t_p^2\right)$ leads to $t_* = \sqrt{\ln\sqrt{P_{\mathrm{in}}/P'_{\mathrm{cr}}}}$. In order to obtain those functions $\mu(t)$ that give rise to stationary solutions with a conserved Gaussian power profile, a standard trust-region method [38] for nonlinear optimization in MatLab is used. This yields a continuum of stationary states, the on-axis intensity profiles of which are depicted in Fig. 3.7a. The propagation constant $\mu(t)$ of the solution is represented by the red curve in Fig. 3.7a. The on-axis profiles feature the same characteristic double-hump temporal structure as in the numerical simulations. A similar analysis on steady-state solutions was carried out earlier in [39, 40], however, with no prediction on pulse break-up. The intensity distribution in the (t, r) plane shown in Fig. 3.7b demonstrates that the plasma nonlinearity acts to defocus the temporal slices around zero delay into a spatial ring as was also observed in the simulations (cf. Fig. 3.6a, c). In summary, evaluation of stationary solutions of (3.6) provides remarkable accurate predictions for the on-axis temporal profile and pulse breaking, occurring in a regime where plasma defocusing balances Kerr self-focusing. Both, the numerical simulations as well as the stationary states calculated directly from the NLSE confirm that the emerging double-hump intensity distributions are defocused around zero delay. These time-slices actually contain the highest optical energy. As the impact of Kerr self-focusing is crucially determined by the optical power rather than intensity (cf. Appendix A), this behavior may be considered counterintuitive. In the following, the position of the local minimum in the double hump structure is scrutinized. This minimum is generally observed to occur when the competing nonlinear effects balance each other at any instant. In a time-dependent variational approach [23, 41], this condition gives rise to the nonlinear integral equation Eq. (3.10) for the on-axis intensity profile $I_0(t)$ of the stationary state. A continuum of solutions of (3.10) is shown in Fig. 3.8a. Quite remarkably, these solutions are in good agreement with the solutions derived directly from the NLSE, showing the characteristic double-hump structure with a minimum around zero delay. In order to calculate the exact position of the minimum we differentiate (3.10) with respect to the retarded time variable t and subsequently set $\partial/\partial t\, I_0(t) = 0$ in the resulting expression. This yields the nonlinear integral equation

$$0 = \dot{G}(t)I_0^2(t) + \frac{\kappa}{(1+N^*)^2}\frac{I_0^{N^*+1}(t)}{P(t)} - 2\kappa N^*\frac{\dot{P}(t)}{I_0(t)}\int\limits_{-\infty}^{t} dt'\, K[t, t', I_0(t), I_0(t')]$$

$$(3.14)$$

with $\kappa = k_0^2 N^* \sigma_{N^*}\rho_0/\pi\rho_c$ and

$$G(t) = \frac{1 - P(t)/P_{cr}}{P^2(t)}$$

and an integral kernel

$$K[t, t', I_0(t), I_0(t')] = \frac{I_0^{N^*+2}(t')}{P^2(t')\left(1 + N^*\frac{I_0(t')P(t)}{I_0(t)P(t')}\right)^3},$$

in which $P_{cr} = \lambda_0^2/(2\pi n_0 n_2)$. Again, the nonlinear integral equation Eq. (3.14) is a generalized Volterra-Urysohn integral equation [24]. Combining a Clenshaw–Curtis quadrature scheme for the integral term of (3.14) with a Chebyshev expansion of the unknown function $I_0(t)$ yields a set of nonlinear equations for the expansion coefficients [26], which are solved utilizing standard algorithms for nonlinear optimization in MatLab [38]. The solution of this equation is depicted by the dotted line in Fig. 3.8a. Moving along this line towards positive times, the local minimum of the

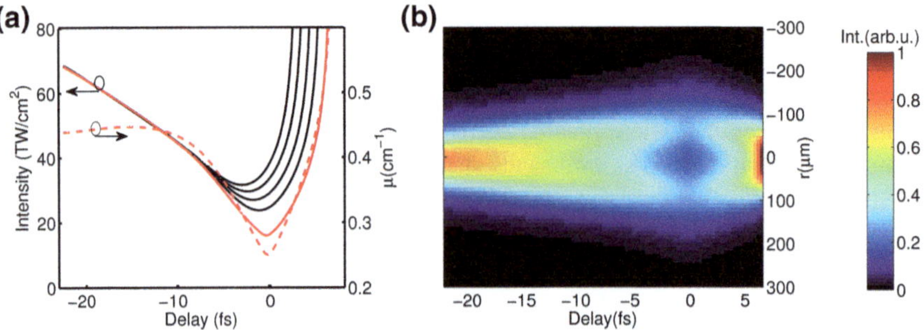

Fig. 3.7 **a** On-axis temporal intensity profiles (*solid lines*) and propagation constant $\mu(t)$ corresponding to the *red curve* (*dashed line*) of steady state solutions, obtained by solving Eq. (3.13), imposing a Gaussian power profile. **b** depicts the spatio-temporal representation of the curve marked in *red*. Nontrivial stationary solutions only exist within the time-window $-t_* \leq t \leq t_*$ $(t_* \approx 22.4\,\text{fs})$. Reprinted from C. Brée et al. [34]. Copyright © 2010 Astro, Ltd.

solution appears more pronounced. This indicates that the pulse splitting mechanism works most effectively in the vicinity of zero delay. This analysis therefore explains the peculiarity of the split preferentially occurring at the instant of maximum power inside the pulse. Although the variational approach provides a good estimate on the exact on-axis temporal profile of the steady states shown in Fig. 3.7a, one can certainly not expect a precise coincidence with the exact solutions, as the variational approach imposes a fixed Gaussian radial shape of the pulse as shown in Fig. 3.8b. In particular, the simplifying assumption of a Gaussian spatial profile neglects the fact that plasma defocusing actually gives rise to the formation of spatial rings. Nevertheless, our analysis corroborates a tendency for self-pinching and pulse break-up. Starting from an independently obtained observation of filamentary pulse-breakup both in numerical simulations and experimental investigations, stationary states of the NLSE coupled to a noninstantaneous plasma response were investigated. The resulting solutions provide a remarkable prediction for the plasma-induced break-up scenario in the strongly ionized filament channel. The quality of the exact solutions compares favorably to stationary solution obtained from Eq. (3.10). In particular, the position of the local minimum separating the individual sub-pulses is directly obtained from a nonlinear integral equation. Both, the exact and the variational approach of deriving stationary solutions to the NLSE corroborates the temporal break-up observed in the numerical simulations and the emergence of local minima of the intensity profile around zero delay. In summary, it can be concluded that with plasma defocusing saturating the optical breakdown, the present assumption of emerging steady state profiles offers deep insight on the dynamical behavior and underlying mechanisms of a physical system that was previously only accessible in detailed numerical simulations.

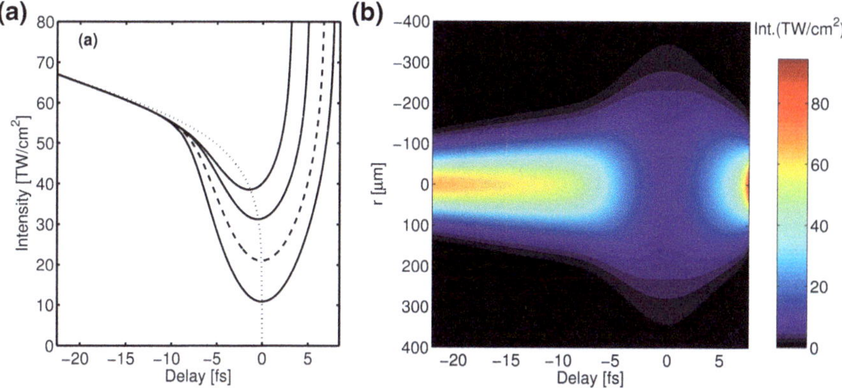

Fig. 3.8 **a** On-axis intensity profile of steady state solutions calculated from a variational approach according to Eq. (3.10). The *dotted line* represents a solution to Eq. (3.14), indicating the position of the local minimum. **b** Depicts the spatio-temporal representation of the curve marked in *red*. Reprinted from C. Brée et al. [34]. Copyright © 2010 Astro, Ltd.

3.3 Cascaded Self-Compression

Besides the so-far discussed filamentation regime leading to self-compression during a split-isolation cycle, filamentation regimes exist which exhibit a richer dynamics in the so-called post-ionization regime following the nonlinear focus [42]. While the first scenario is characterized by a single, strongly-ionized regime followed by a weakly-ionized, sub-diffractive channel, the latter exhibits multiple refocusing events in the postionization regime leading to multiple temporal splittings of the pulse. The appearance of refocusing events and multiple temporal splittings in nonlinear filament propagation is closely related to the occurrence of modulational instabilities. These instabilities are a characteristic feature of the nonlinear propagation of waves. Prototypical examples are the Benjamin–Feir instability [43] of deep-water waves and the azimuthal modulational instability of spatial solitons of the nonlinear Schrödinger equation in optics [44]. Similar phenomena have been reported to occur in Bose–Einstein condensates [45], in plasma physics [46, 47] and in the propagation of short laser pulses [48–50]. In self-generated optical filaments, temporal break-ups serve to actively compress femtosecond laser pulses [41, 51]. Recently, there has been revived interest in such phenomena as they can give rise to an unusual increase of pulse amplitude or concentration of energy and to the appearance of the so-called rogue waves [52, 53]. The probability for the appearance of these rare events rapidly decreases with their amplitude. As the physical systems are deterministic, perfect control of the input wave should, in principle, enable an arbitrary increase of wave amplitude within the system's limitations. However, exploitation of rogue wave phenomena [54] or other highly nonlinear scenarios for the generation of a desired pulse shape is technically limited by the feasibility of control over the input wave. Noise on the input waveform therefore impedes pulse compression at a certain point.

In most systems, a fundamental limitation arises due to quantum noise [49]. In the current section, a new approach for exploiting rare events in a highly nonlinear system is presented, cascading the process while at the same time limiting the underlying nonlinearity in every step. The latter measure maintains control over the output wave when exploiting such events, e.g., for waveform compression. This cascaded waveform control is illustrated by the propagation of short pulses in a self-generated filament that are suitably described by the nonlinear Schrödinger equation [41, 51]. In this system, pulse compression factors of the order of 3–5 have been previously discussed [11, 15, 18, 55, 56] in single-compression cycles. Quite remarkably, the compression factor of each individual process remains nearly conserved in double self-compression, enabling in total nearly twelve fold compression [57].

For an investigation of this cascaded compression mechanism, numerical simulations of the generalized Nonlinear Schrödinger Equation (2.25) are conducted, modeling the propagation of intense femtosecond pulses in argon [58, 59]. Based on experimental parameters discussed below, input pulses containing an energy of 2.5 mJ at 800 nm and initial beam waist $w_0 = 2.5$ mm are assumed. The input pulse duration was taken as $t_{FWHM} = 120$ fs. The pulses are focused by an $f = 1.5$ m lens into a noble gas. To identify the small parameter range giving rise to compressed output waveforms on the axis of the filament, a parameter scan is performed by varying the gas pressure in a range from 100 to 120 kPa, at otherwise fixed input parameters, see Fig. (3.9). At a pressure $p = 106$ kPa, the simulations predict plasma-dominated dynamics in a relatively short nonlinear focal zone succeeded by a 1 m long self-generated channel, in which plasma formation is virtually absent. This qualitatively corresponds to the numerical scenario analyzed in Sect. 3.1. In the plasma-free zone, Kerr self-focusing effectively balances linear diffraction, see Fig. 3.9a. This figure clearly reveals how a splitting event at $z = 1.4$ m close to the linear focus position merges into formation of one isolated and shorter pulse. The splitting initially produces two pulses, one at $t = -100$ fs and a second one at $t = +60$ fs. At $z = 1.6$ m, each of these subpulses is roughly 40 fs wide, which is a natural consequence of the split. Upon further propagation ($z = 1.7$ m), the pulse at negative delays dies out quickly, leaving only one isolated and shortened pulse behind. This prototypical split-isolation cycle has already been discussed in [15, 18, 34] as the origin of on-axis pulse self-compression [11]. After the split-isolation cycle, plasma generation has effectively ceased, such that pulse shaping in the elongated channel at $z > 1.7$ m is now dominated by an interplay between Kerr-type self-refraction and linear optical effects. Notably, self-focusing compensates diffractive optical effects, giving rise to a sub-diffractive nature of this final nonlinear propagation stage as also observed by Faccio et al. [60].

Increasing the pressure to 109 kPa, the delicate balance in the sub-diffractive channel behind the strongly ionized zone is disturbed by a slight increase of Kerr nonlinearity. This increase triggers a refocusing event 0.5 m behind the first nonlinear focus, and a second strongly ionized zone evolves (Fig. 3.9b, c). In fact, in Ref. [61] it has been shown shown that already 0.5 % amplitude noise on the input waveform yields shot-to-shot intensity fluctuations in the second focus of about 50 %.

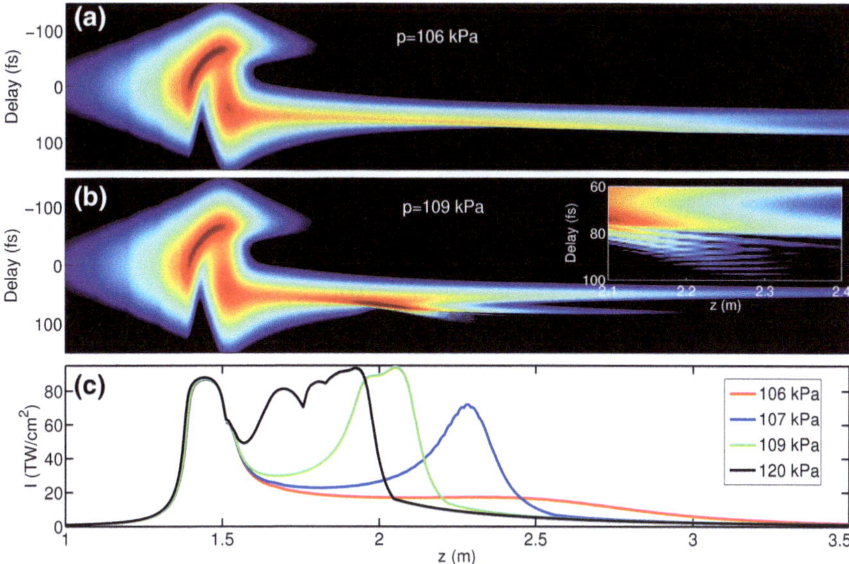

Fig. 3.9 **a** Evolution of on-axis intensity profile along z for pulse self-compression into a sub-diffractive channel in argon, $p = 106$ kPa. **b** Corresponding evolution for the double self-compression scenario at $p = 109$ kPa. Inset: close-up on the pulse break-up in the second focus, accompanied by shock wave formation. **c** Evolution of on-axis peak intensity for pressure from 106 to 120 kPa. Reprinted from C. Brée et al. [57]. Copyright © 2010 IOP Publishing, Ltd.

While refocusing, the pulse experiences a second split-isolation cycle that shows superficially the same behavior as the first one, i.e., the surviving pulse from the first cycle splits into two at $z = 2.2$ m. In contrast to the first cycle, the trailing pulse dies out at $t \approx 80$ fs, leaving only one isolated and yet again shortened pulse at $t \approx 50$ fs behind. In the subsequent nonlinear propagation inside the channel, the pulse reaches a minimum duration of 16.4 fs at $z = 2.5$ m. Further increasing the pressure to $p = 120$ kPa, pulses with a minimum duration of 10.9 fs emerge after the second focus. This nearly twelve fold compression chiefly goes back to the two split-isolation cycles. Such a strong compression effect has neither been observed in previous experimental [11, 55, 56] nor theoretical studies [15, 18]. The emergence of the refocusing event must not be confused with focusing-defocusing cycles [17] that occur on significantly shorter length scales of ≈20 cm whereas repetition of the split-isolation cycle is only observed with a distance >50 cm between the events. Apart from the different length scales, our simulation indicate a pronounced intensity drop and a resulting cease of plasma formation between the two cycles (Fig. 3.9c), which further suggests a conceptual difference to the much milder focusing-defocusing cycles previously reported.

Despite the apparently identical effect on pulse duration, collapse saturation in the two foci is accomplished by different physical effects. In the first nonlinear focus ($z = 1.5$ m), plasma defocusing and related dissipative terms clamp the intensity whereas

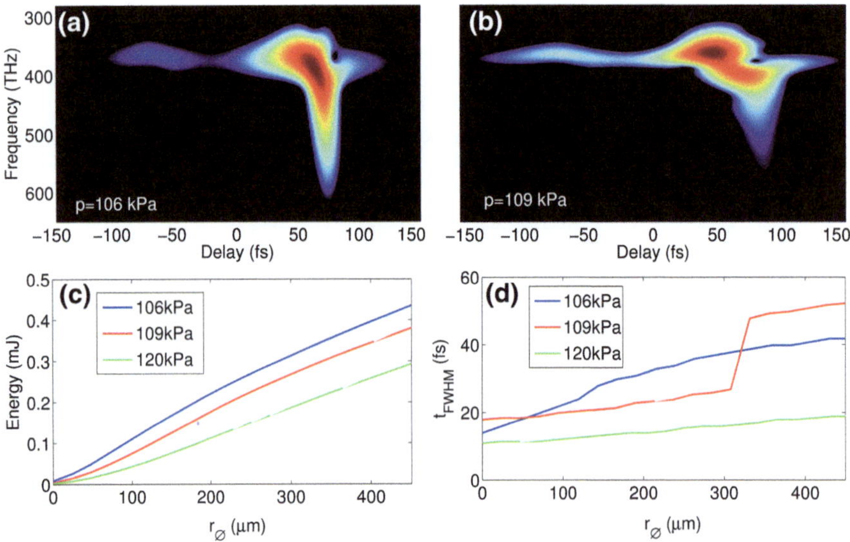

Fig. 3.10 On-axis XFROG spectrograms **a** of the optical field emerging from the sub-diffractive channel regime of Fig. 3.9a, $z = 2.45$ m, $p = 106$ kPa, and **b** after the second pulse breaking at $z = 2.45$ m, $p = 109$ kPa (two-foci regime). **c** Energy of the output pulse at $z = 2.5$ m transmitted through an aperture of radius r_O. (d) Temporal duration (FWHM) of transmitted power profiles. Reprinted from C. Brée et al. [57]. Copyright © 2010 IOP Publishing, Ltd.

temporal effects, in particular dispersion, take over this role in the second focus ($z = 2$ m). Indeed, neglecting the plasma response for $z > 1.75$ m in the simulation at $p = 107$ kPa (Fig. 3.9c, blue line) leads to a nearly unchanged dynamical behavior for the second compression stage. Similar plasmaless refocusing events have been discussed in Refs. [42, 62]. With increasing pressure ($p \geq 1.09$ bar), however, plasma becomes again essential for preventing spatial wave collapse, while dispersion dominates temporal dynamics by exchanging power between different pulse time slices. The generation of dispersive shock waves in the trailing edge of the pulse (Fig. 3.9b, inset) during the second splitting event further underlines the strong impact of dispersion and self-steepening.

In order to further analyze spectro-temporal signatures of the pulse-shaping action during the refocusing event, XFROG spectrograms were computed from the simulated on-axis data, see Fig. 3.10. XFROG spectrograms are a convenient way to analyze characteristic deviations from a spectrally and temporally homogeneous energy distribution inside the pulse, which are also directly measurable [63], see also Appendix C. In highly nonlinear scenarios, these spectrograms have previously elucidated the mechanisms behind supercontinuum generation in photonic crystal fibers [64] and filaments [15]. In the single self-compression regime, these representations exhibit a characteristic shape that are most suitably described as the mirror image of the Greek letter Γ (Fig. 3.10a), as has already been discussed in [15]. A short pulse duration is intimately connected to a vanishing slant of the vertical bar of the Γ. The extension

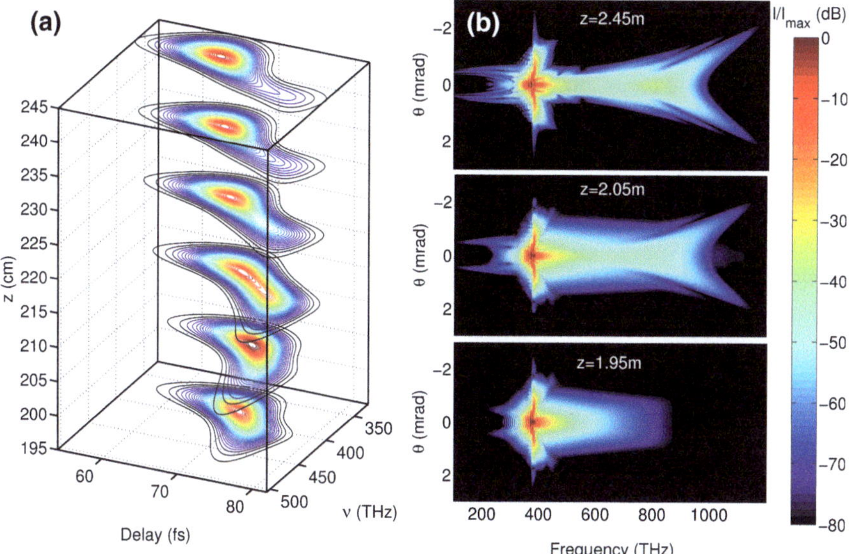

Fig. 3.11 **a** Evolution of XFROG spectrogram along z during propagation through the second focus. **b** Corresponding angularly resolved far-field spectra. Reprinted from C. Brée et al. [57]. Copyright © 2010 IOP Publishing, Ltd.

of this section towards the blue is a measure of the asymmetric nonlinear spectral broadening effects, mainly caused by self-steepening [65]. The appearance of pronounced horizontal structures along the cap section of the Γ, in contrast, is connected to the suppression of the leading pulse in the split-isolation cycle, i.e., pulse contrast. If a second split-isolation cycle appears, the spectro-temporal pattern of the pulses changes in a characteristic way, see Fig. 3.10b. Remnants of the suppressed pulse after the second split-isolation cycle now appear as a blue trailing pedestal of the spectrogram, i.e., point symmetric to the red leading pedestal appearing after the first cycle, with a shape that we will refer to as Q-shape in the following. The broadening effect appears as a spectral red-shift along the vertical axis in Fig. 3.10b. As filamentary self-compression is typically restricted to a small region around the optical axis, for the simulated output pulses at $z = 2.5\,\mathrm{m}$, the power profile transmitted through an aperture of radius r_0 defined by $P_0(t) = 2\pi \int_0^{r_0} dr\, r I(t, r)$ was calculated, where $I(t, r)$ defines the spatiotemporal intensity distribution of the laser field. The transmitted energy and averaged pulse duration versus aperture radius r_0 are shown in Fig. 3.10c, d, respectively. This clearly shows that, for the chosen initial conditions, double self-compression is superior to the single-focus scenario for two reasons. First, the pulse duration (green curve) increases less rapidly with r_0 for the double-compression scenario at $120\,\mathrm{kPa}$. Second, only for this scenario, energetic $0.3\,\mathrm{mJ}$ output pulses at sub-$20\,\mathrm{fs}$ duration can be obtained.

Figure 3.11a shows a more detailed view of the transition from inverse Γ to Q-shape, with a zoomed-in set of spectrograms computed in the range from $z = 195$

to 245 cm. In the initial spectrograms in this sequence, the typical Γ shape appears with the vertical bar extending into the blue spectral range. During the approach towards the second focus, however, the spectral extension into the blue reduces, with a red shift appearing shortly afterward. Emergence of the red-shift comes along with formation of a blue trailing pedestal, which is ultimately a remnant of the blue wing of the inverse Γ shape. Figure 3.11b additionally shows a view of angularly resolved spectra [60] during this transition phase. These structures exhibit a markedly different behavior in the blue and the red wing of the pulse. This behavior is related to the strong influence of self-steepening, which causes the blue shift in the trailing part of the pulse. In fact, the modulational instability occurring in self-focusing media with normal group-velocity dispersion reshapes this part of the pulse into a characteristic X-shaped spatio-spectral pattern [66]. The X-waves are known to emerge in regimes where normal group velocity dispersion saturates the optical collapse due to Kerr self-focusing. Mathematically, they originate from exponentially amplified noise on a cw-beam with the radial shape of a Townes soliton, being a stationary, yet unstable, solution of the simplified Nonlinear Schrödinger equation

$$\partial_z \mathcal{E} = \frac{i}{2k_0}\Delta_\perp \mathcal{E} - i\frac{\beta_2}{2}\frac{\partial^2}{\partial t^2}\mathcal{E} + i\frac{\omega_0}{c}n_2|\mathcal{E}|^2\mathcal{E} \qquad (3.15)$$

with GVD coefficient $\beta_2 > 0$. In the spatio-temporal domain, this instability has also been shown to be responsible for the observed temporal splitting and the emergence of hyperbolic shock waves [67, 68]. Remarkably, those dispersion dominated dynamics are still observable in the pressure regime above 109 kPa, where plasma defocusing is already essential for wave-collapse arrest. The apparent red-shift of the spectra of the optical fields emerging from the 2-foci regime can thus be ascribed to both, self-phase modulation in the leading edge of the pulse during the refocusing stage and to angular dispersion of the blue spectral content of the pulse into a spatial reservoir due to the generation of the shock wave. Figure 3.12a shows the shock-wave in the spatiotemporal domain, exhibiting a clear temporal asymmetry, with shock-waves only appearing in the trailing part of the pulse at $t \approx 80$ fs. This characteristic asymmetry shows up in the few-cycle regime, where self-steepening and space-time focusing have to be taken account in the model by including the operator T and T^{-1}. In fact, presupposing $T = 1$, theory predicts temporally symmetric shock-waves both in the leading and trailing edge of the pulse, an observation that has also been done in the context of nonlinear fiber optics [69], where shock-waves are generated both in the leading and trailing parts of the pulse due to the interplay of normal GVD and self-phase modulation. In the trailing part of the pulse, Fig. 3.12a exhibits the presence of a spatial ring approximately localized at $t = 70$ fs and $r = 600\,\mu$m. In order to analyze this phenomenon in more detail, Fig. 3.12b displays a tomographic representation of the laser pulse, showing spectrogram representations of the pulse for different radial distances from the optical axis. The spatial ring shown in Fig. 3.12a is also clearly visible in the XFROG tomographic representation of Fig. 3.12b, with a blue-shift of approximately 20 THz w.r.t. the carrier frequency of 375 THz. In fact, the emergence of a spatial ring with anti-Stokes shifted wavelengths around the

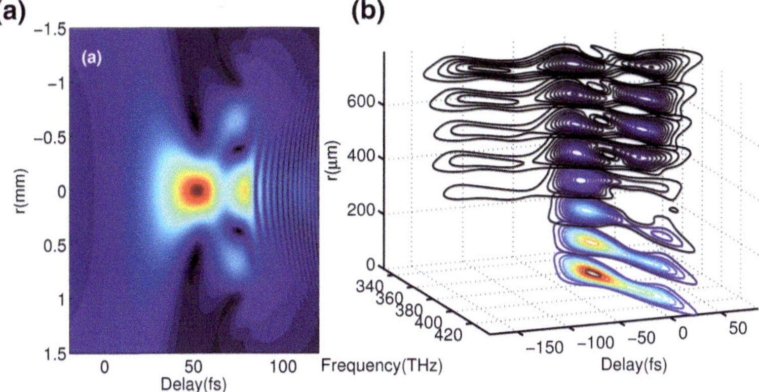

Fig. 3.12 **a** Intensity distribution in the (r, t)-plane of a filamentary pulse at $z = 2.5$ m and $p = 1.09$ kPa. **b** Tomographic XFROG representation of same pulse for different distance r from optical axis

white-light core of the filament is a well-known phenomenon and is referred to as conical emission (CE). Recently, the CE has been shown to be related to X-waves, and, beyond that, has been described in terms of a description similar to that of Cerenkov-radiation [70].

3.3.1 Experimental Evidence of Cascaded Self-Compression

For experimental verification of the double self-compression, a 45 fs regenerative Ti:sapphire amplifier system with a pulse energy of 5 mJ was employed. The laser pulse energy has been carefully attenuated by means of an adjustable diaphragm and focused with an $f = 1.5$ m lens to generate a single filament in air. A second diaphragm after the filament served to isolate the core of the filament. After suitable attenuation, the temporal structure in the filament core was analyzed with spectral phase interferometry for direct electric-field reconstruction (SPIDER, [71–73]). The SPIDER method delivers the spectral phase, which can be combined with an independently measured spectrum to reconstruct the complex-valued field envelope in the spectral or temporal domain. Moreover, this information also suffices to directly reconstruct XFROG spectrograms from experimental data.

Except for the fact that no gas cell was necessary, this setup widely resembles the one in [11]. Adjusting the input diaphragm, a regime could be found that displayed a single filament with two clearly separated strongly ionized zones that were separated by about 30–40 cm. With these short input pulses, the simulations indicate that we can at best expect about threefold compression. It may appear intriguing to suggest dispersive stretching of the 45 fs pulses to 120 fs duration in order to demonstrate the full compression potential. Yet, these chirped pulses would already exhibit a

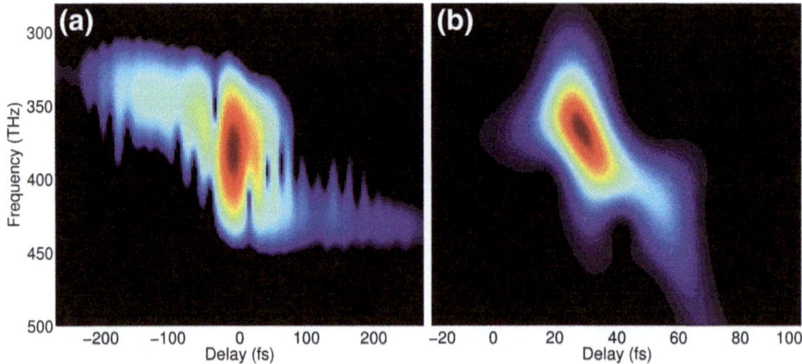

Fig. 3.13 **a** XFROG trace of an output pulse after double self-compression in air, obtained from measured SPIDER data. **b** Numerically obtained Q-shaped XFROG trace at the exit of the air filament after double self-compression, $z = 3$ m. Reprinted from C. Brée et al. [57]. Copyright © 2010 IOP Publishing, Ltd.

much wider bandwidth than Fourier-limited 120 fs pulses, and compression could also stem from linear optical effects. Therefore the short 45 fs pulses delivered by the laser source were used as the input directly.

From the measured SPIDER data, the XFROG spectrogram shown in Fig. 3.13a was reconstructed. This spectrogram shows features previously discussed for the single-focus and the double-focus regime in argon, cf. Fig. 3.10a, b, respectively. From the former, a temporally stretched leading pedestal is discernible, which is generally quite typical for self-compression [15]. In addition to previous experimental findings, however, a clearly visible trailing blue pedestal appears. A Q-shaped spectrogram thus forms, which characterizes a second split-isolation cycle. Such a structure has not been reported in literature yet. It is striking that this feature appears temporally less stretched than the leading red pedestal from the first split-isolation cycle, which corroborates less exposure to linear and nonlinear pulse shaping effects. Therefore the experimental findings appear to be highly compatible with the causal sequence of events predicted by numerical simulation. This finding also suggests that the second split-isolation cycle is being caused by a different mechanism than the first one, causing pedestal formation at opposing spectral and temporal edges of the main pulse. To compare these experimental findings with theoretical predictions, a delayed Kerr-nonlinearity was included in the model equations for pulse propagation in molecular air and additional numerical simulations were performed, with 2.5 mJ Gaussian input pulses $w_0 = 3.5$ mm, $t_{\mathrm{FWHM}} = 45$ fs. These initial conditions match the experimental input pulse parameters as close as possible. The numerical simulation shows two distinct ionization zones and a characteristic Q-shaped XFROG spectrogram (Fig. 3.13b) emerging after the refocusing stage and corresponding split-isolation cycle (Fig. 3.14a). Thus, the numerical data reproduce the characteristical features of the measured pulses, including redshifted leading and blueshifted trailing pedestals, as also observed in numerical simulations of double self-compression in argon. In addition, Fig. 3.14b shows spectra from experiment and theory. Both sim-

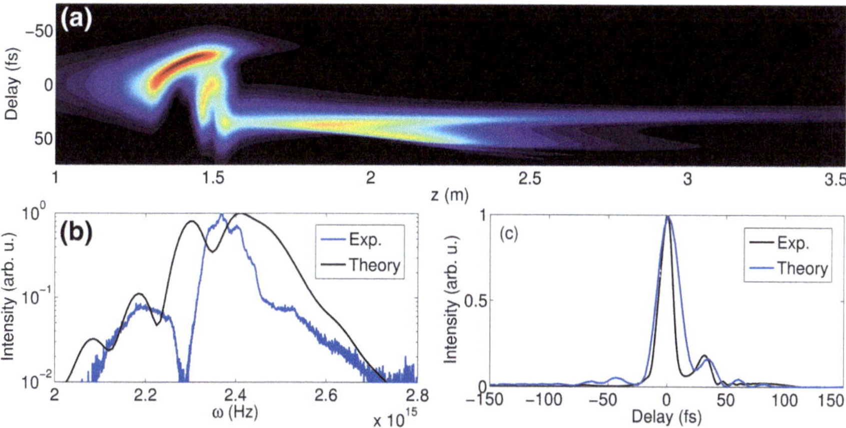

Fig. 3.14 **a** Evolution of on-axis temporal intensity along z in a numerically simulated filament in air, exhibiting refocusing stage and double splitting events. **b** On-axis spectra in numerical simulation and experiment. **c** On-axis temporal intensity profile from SPIDER measurement (*blue curve*) versus on-axis profile at $z = 3.5$ m obtained from numerical simulations (*black curve*). Reprinted from C. Brée et al. [57]. Copyright © 2010 IOP Publishing, Ltd.

ulated and experimentally recorded spectra exhibit a pronounced redshift, which, according to our previous discussion, emerges due to spatio-spectral reshaping of the pulse during the refocusing stage. In Fig. 3.14c a comparison is shown between on-axis temporal profiles of measured and simulated pulses. The measured pulse exhibits a duration $t_{\text{FWHM}} = 22$ fs, while the simulated pulse has $t_{\text{FWHM}} = 14$ fs on-axis.

The cascaded compression scenario is not an isolated phenomenon, but can be obtained for a range of input pulse parameters and gas species, which sets it apart from a highly optimized single-compression scenario. Assuming a different gas, e.g. krypton as the nonlinear medium [58, 59], for a demonstration of the universality of this mechanism the parameter range of input pulse energy and peak power was scanned in numerical simulations for appearance of this phenomenon. Beam waist and temporal duration were fixed at $w_0 = 5$ mm and $t_{\text{FWHM}} = 120$ fs, respectively. The observed pulse shortening as a function of input energy and system nonlinearity (peak power normalized to P_{cr}) is depicted in Fig. 3.15, with iso-pressure lines shown in white. The dashed line, roughly collinear with the 100 kPa pressure line, marks the lower limit of double self-compression. From this picture, the capability of the cascaded self-compression becomes immediately clear, giving rise to up to twelve fold compression. Compression ratios above 10 are localized in the region of double self-compression and can already be observed at powers exceeding the critical power by a factor of only three. Our scan also reveals examples for threefold cascading of the split-isolation cycle, yet with imperfect isolation in the last cycle. Generally, for pressures exceeding 160 kPa an increased tendency for such undesired multiple temporal splits is observed. Importantly, cascaded self-compression fills a

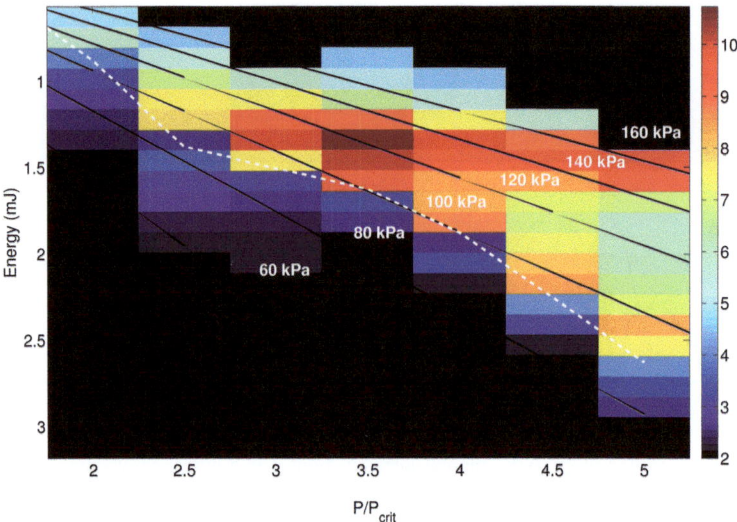

Fig. 3.15 Pulse compression ratio in a krypton filament for various initial conditions in the Energy versus P/P_{cr} plane. The *solid lines* correspond to lines of equal pressure. The *dashed line* separates subdiffrative channel regimes (*below*) from double self-compression regimes (*above*). Reprinted from C. Brée et al. [57]. Copyright © 2010 IOP Publishing, Ltd.

considerable fraction of the parameter space mapped out in Fig. 3.15. This sets it apart from sparsely represented rogue-wave-like events [53].

The conducted numerical investigations and experimental studies pinpoint an alternative approach toward efficient exploitation and control of highly nonlinear wave shaping mechanisms. Rather than trying to confine input parameters in an increasingly narrow range, it appears much more promising to relax these constraints in order to avoid that input noise strongly affects the output waveform. It was demonstrated that physical systems exist that allow for cascaded application of the waveform shaping effect, e.g., in order to compress optical pulses or to concentrate energy. While this effect certainly also narrows the input parameter space, it is minor as compared to immediate rogue wave control that exhibits a rapidly imploding parameter space with increasing amplitude [52]. The cascaded compression method presented here therefore opens a perspective not only for optical pulse compression but for exploitation of waveform control in a wide range of similar highly nonlinear physical scenarios.

3.4 Temporal Self-Restoration in Femtosecond Filaments

Spatiotemporal soliton solutions [74] to the Maxwell equations in a nonlinear medium are an intriguing concept, as these hypothetical objects may serve to transfer electromagnetic energy over distances widely exceeding the Rayleigh range of linearly diffracting laser beams. Indeed, with the discovery of filamentary light propagation, it was observed for the first time that nonlinear optical effects are able to balance diffraction, giving rise to extended, sub-diffractive channels of light maintaining a high fluence along their longitudinal extension. This unexpected property has led the authors of [75] to conclude that the physics of femtosecond filaments is allegeable in terms of spatiotemporal optical solitons, an assumption that did not remain undisputed [76], however. Indeed, contrary to the idea of spatiotemporal solitons, as it was pointed out in the previous sections, filamentary propagation of light is a highly dynamical process, with a recurrence of focusing/defocusing cycles maintaining the illusion of stationarity [28]. Nevertheless, it is frequently observed that filaments are very robust objects and exhibit self-restoration capabilities of their spatial profile when hitting, e.g., small water droplets [77]. These properties are rather typical for solitary solutions of nonlinear propagation equations, and it is in fact possible to show that a time average of the propagation equation governing the filament dynamics admits spatial soliton solutions [77, 78]. Moreover, it has recently been shown for the first time that filamentary light pulses admit self-restoration of their temporal profile when exiting a gas-filled chamber through a silica window, a setup typically used in filament self-compression experiments. That is to say, the detailed theoretical analysis of [79, 80] shows that the self-compressed laser pulse experiences a dramatic change of its temporal profile as it traverses the silica window, a behavior which is not explicable by linear theory. In fact, for a few-cycle pulse with initial duration $t_{\mathrm{FWHM}} = 10\,\mathrm{fs}$ traversing a silica sample with thickness $\Delta z = 500\,\mu\mathrm{m}$, according to Eq. (2.68), only a moderate temporal broadening due to the GVD of silica ($\beta_2 = 370\,\mathrm{fs}^2/\mathrm{cm}$) to $t_{\mathrm{FWHM}} = 11.2\,\mathrm{fs}$ is expected. In contrast, theoretical work [80] reports temporal stretching in a 0.5 mm thick silica window from initially 13 to 28 fs and $>30\,\mathrm{fs}$, respectively. Surprisingly, it is shown in the latter work that the pulse is able to recompress after it leaves the silica window. This surprising behavior, i.e. the dramatic temporal stretching inside the exit window and the subsequent recompression, was analyzed in more detail in [80]. Figure 3.16 taken from [79] shows the simulated evolution of the on-axis intensity profile for two different pulses directly after they leave the exit window and continue propagation in air. The two configurations are distinguished by the position of the exit window, where in the upper left case, the exit window is located closer to the filament. Both upper panels clearly show the strong temporal broadening of the on-axis profile directly after the exit window. Further on, it is evident that temporal self-restoration is accompanied by a refocusing stage. It has been shown in [79] that this refocusing stage emerges from a spatial phase curvature resulting from spatial SPM which acts like a focusing lens.

Thus, by robustness of their spatiotemporal profile, filamentary light pulses exemplify properties typical for spatiotemporal soliton solutions, although the underlying dynamical equations generally do not admit unconditionally stable soliton solution in space and time. These theoretical predictions by Bergé et al. provided a motivation for an experimental study of the influence of the exit window on filamentary

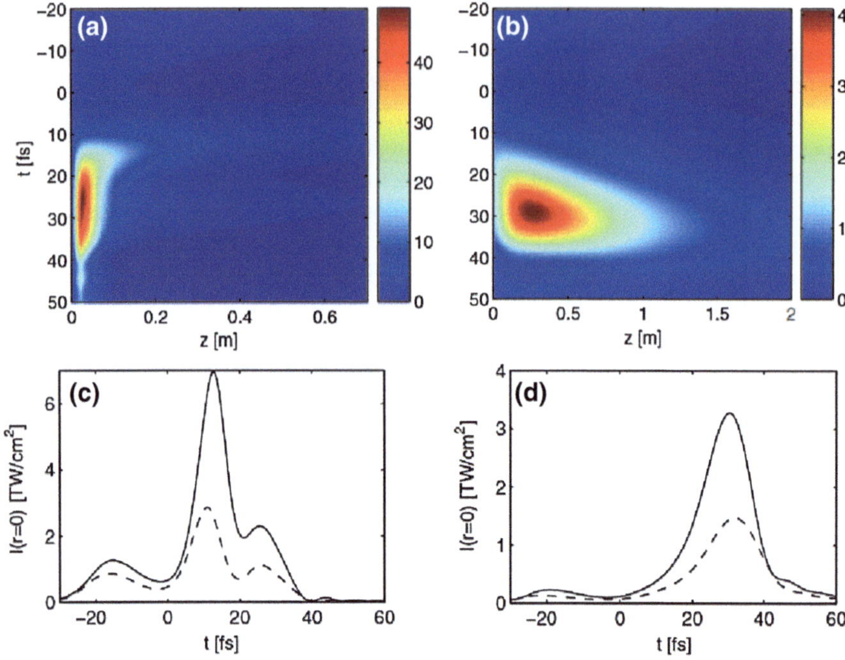

Fig. 3.16 Main features of temporal self-restoration as numerically predicted in [79]. The figure is taken from Ref. [79]. Reprinted from L. Berge et al. L.2008. Copyright © 2008 American Physical Society

propagation [81]. However, due to the high intensities involved, the pulse profiles directly before and after the exit window elude any measurement attempt, such that the details of the evolution leading to self-recompression are inaccessible experimentally. Nevertheless, it is possible to obtain indirect evidence of the influence of the exit window, as will be demonstrated in the following.

3.4.1 Experimental Prerequisites

The employed experimental setup is shown in Fig. 3.17. The laser source is a Ti:sapphire regenerative amplifier delivering 45 fs pulses at 5 mJ and beam waist $w_0 = 9$ mm. Focusing the pulsed laser beam with a R $= 3$ m parabolic mirror into an argon-filled gas cell at atmospheric pressure, an approximately 40 cm long filament is formed, with a reddish violet fluorescent trail starting at about 10 cm before the geometrical focus. In order to achieve a single stable filament, a central part of the beam is selected by an adjustable aperture (D1). The optical power transmitted through this first aperture was measured as 1.4 W, which at a repetition rate of 1 kHz corresponds to a pulse energy of 1.4 mJ and to a peak pulse power of 30 GW. This

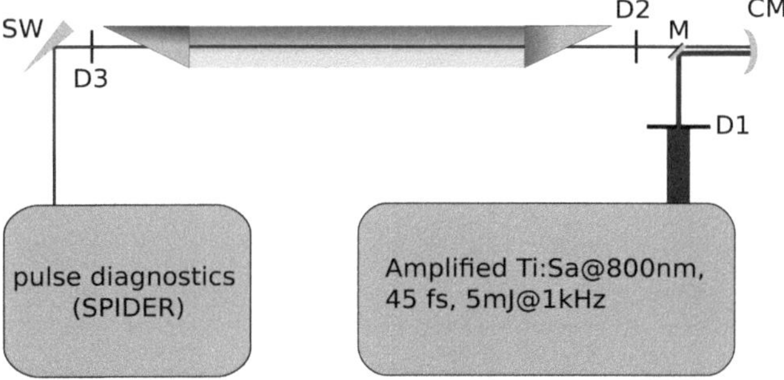

Fig. 3.17 Experimental setup: adjustable diaphragms D1, D2, D3, curved mirror CM ($R = 3.0$ m), mirror M, silica wedge SW. Pulsed femtosecond laser radiation generated within a Ti:sapphire regenerative amplifier at a repetition rate of 1 kHz, emitting 45 fs, 5 mJ pulses at 800 nm center wavelength

equals three critical powers for self-focusing in argon at 800 nm. At the output of the gas-cell, a pulse energy of 1.2 mJ was measured. The beam enters and leaves the gas cell by traversing 500 μm thick Brewster-angled silica windows, which were later replaced by 50 μm thick polypropylene foils. As filamentary self-compression dynamics is most pronounced in the spatial center of the beam, aperture D3 was carefully matched to select the white-light core of the beam, which contains a pulse energy of 0.6 mJ. For diagnosis of the filament cell output, a simple spectrograph and spectral phase interferometry for direct electric-field reconstruction (SPIDER, [71–73, 82]) were used. A brief overview about the SPIDER interferometric technique is given in Appendix B. In order to avoid damage in the nonlinear crystal of the SPIDER setup, the beam was attenuated using the front reflex off a silica wedge (SW). As the subsequent experiments strongly rely on the accuracy of the SPIDER data, the statistical fluctuations of the reconstructed output pulse shapes from the self-generated filament were carefully analyzed in a first step.

Figure 3.18a shows a recorded interferogram (solid line) and the standard deviation (dashed line) of the reconstructed group delay, calculated from a set of approximately 50 measured interferograms. Here the SPIDER trace is detected at the second harmonic of the input wavelength, i.e., at about 750 THz, since the SPIDER method relies on the generation of two spectrally sheared, upconverted copies of the pulse by sum frequency generation in a $\chi^{(2)}$-medium. The corresponding spectrum and integrated spectral phase, averaged over the available data, are shown in Fig. 3.18b. In the time domain, Fig. 3.18c shows the temporal intensity profile reconstructed from the averaged phase (strong solid line). The measured pulse clearly exhibits the well-known characteristic asymmetry [11, 15, 65] of filamentary light bullets, with a duration (FWHM) of 18 fs. In addition, the pulse profiles corresponding to the individually measured interferograms (light gray lines) are displayed in the same figure. These measurements show that the modulation depth of the measured SPIDER traces is high enough to yield sufficient accuracy over the relevant spectral

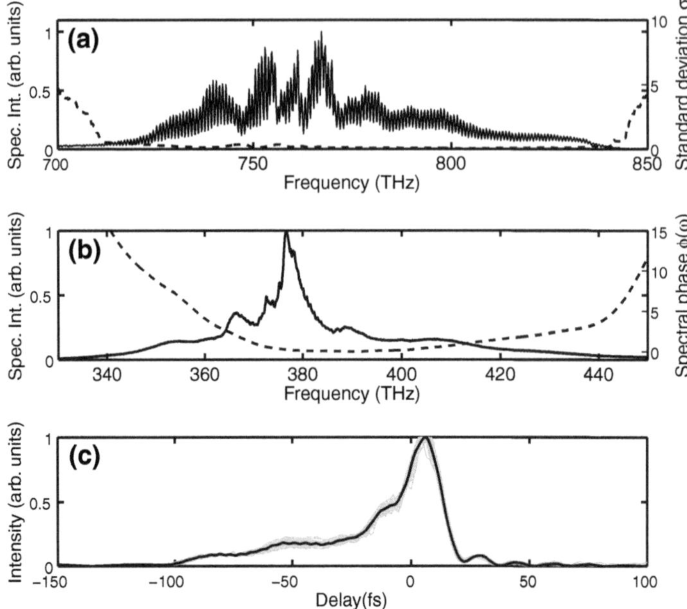

Fig. 3.18 **a** Measured SPIDER interferogram versus frequency $\nu = \omega/2\pi$ (*solid line*) and standard deviation of spectral phases (*dashed line*) retrieved from a set of $\approx$50 interferograms. **b** Spectrum of a self-compressed pulse emerging from a filament (*solid line*). *Dashed line* shows the averaged spectral phase. Concave phases indicate excess normal dispersion. **c** Time domain representation of the measured pulse. The *black line* shows the reconstruction using the averaged spectral phase. The *light gray lines* show pulse profiles reconstructed from each measured interferogram. Reprinted from C. Brée et al. [81]. Copyright © 2008 American Physical Society

range, in particular in the extended blue wing of the spectrum. For the pulse shown, this gives rise to an error in pulse duration not exceeding ± 1 fs.

3.4.2 Experiment 1: Variation of Window Position

In a first experiment the position of the entire argon cell is varied along the optical axis. In particular, this measure varies the distance between the plasma column and the output window, but leaves the linear dispersion of the system unaffected. The window position Δz is defined as the distance between the inner surface of the silica output window of the cell and the position of the geometrical focus. The quantity Δz is varied between $\Delta z = 103$ and 111 cm, the maximum variation allowed due to practical constraints in the experimental setup. The results of the measurements are summarized in Fig. 3.19. In Fig. 3.19a, the measured spectra at minimum and maximum Δz are shown. This comparison clearly reveals that a decrease of Δz comes along with an elevated red shoulder of the spectrum, while the blue shoulder generated by self-phase modulation and self-steepening in the trailing part of the pulse appears suppressed. In fact, an analogous behavior was observed in the numerical simulations of Ref. [80]. The SPIDER setup determines the spectral phase $\Phi(\omega)$ up to an unknown global offset and an arbitrary linear dependence. However, the former only corresponds to a global phase factor while the latter translates the entire pulse in the time domain. Therefore, given that one is only interested in the temporal pulse shape and the phase, all relevant information is encoded in the group velocity dispersion (GDD) defined as $D_2(\omega) = \partial^2\Phi(\omega)/\partial\omega^2$. Consequently, instead of the measured spectral phase, only the corresponding GDD is considered in the following. Figure 3.19b displays the dependence of the measured GDD on the window-filament distance Δz in a frequency range from 340 to 420 THz. This shows that the GDD exhibits strong fluctuations, especially in the red spectral range, i.e., below a carrier frequency of 375 THz. In fact, both the smaller absolute values and the weak variation of the GDD on the blue side of the spectrum are well expected from the asymmetric spectral broadening due to self-steepening which becomes increasingly relevant in the few-cycle domain. Self-steepening generates new blue spectral content in the trailing part of the pulse and is thus strongly localized in the time domain [65]. This is also evident from the Γ-shaped XFROG spectrograms discussed in the previous section. In the frequency domain, this strong localization is evidenced by a nearly flat spectral phase in the blue part of the spectrograms, a feature which has been verified both theoretically and experimentally several times [11, 15, 65]. Reconstructing temporal pulse profiles from measured spectra and phases, as shown in Fig. 3.19c, the pulse duration is constant within experimental precision for large window-filament distances $\Delta z > 109$ cm. However, a reduction of Δz results in an increasing pulse duration. At $\Delta z \leq 106$ cm, only poor compression ratios are observed, with pulse durations exceeding 30 fs. In the following, the functional

$$V^{\mathrm{GDD}}(\omega_1, \omega_2) = \int_{\omega_1}^{\omega_2} \left| \frac{dD_2(\omega)}{d\omega} \right| d\omega \qquad (3.16)$$

measures the total variation of the GDD between $\omega_1 = 2\pi \times 340$ THz and $\omega_2 = 2\pi \times 425$ THz. In Fig. 3.19d, $V^{\mathrm{GDD}}(\omega_1, \omega_2)$ is plotted against Δz. From $\Delta z = 103$ to 109 cm, the variation roughly correlates with the pulse duration, with the exception of $\Delta z > 109$ cm, where it increases while the pulse duration remains nearly constant. These measurements clearly show that the compression ratio strongly depends on the position of the exit window. In fact, an unsuitable choice of the window position can render filamentary self-compression unobservable. However, in order to provide evidence for the self-restoration mechanism of [79, 80], it is clearly necessary to supplement the previous measurements with experimental data of the self-compression efficiency **inside** the argon cell, which is, of course, impossible to achieve due to the high intensity within the filament. Instead, output pulses from a windowless argon cell, as described in the next section, provide indirect evidence

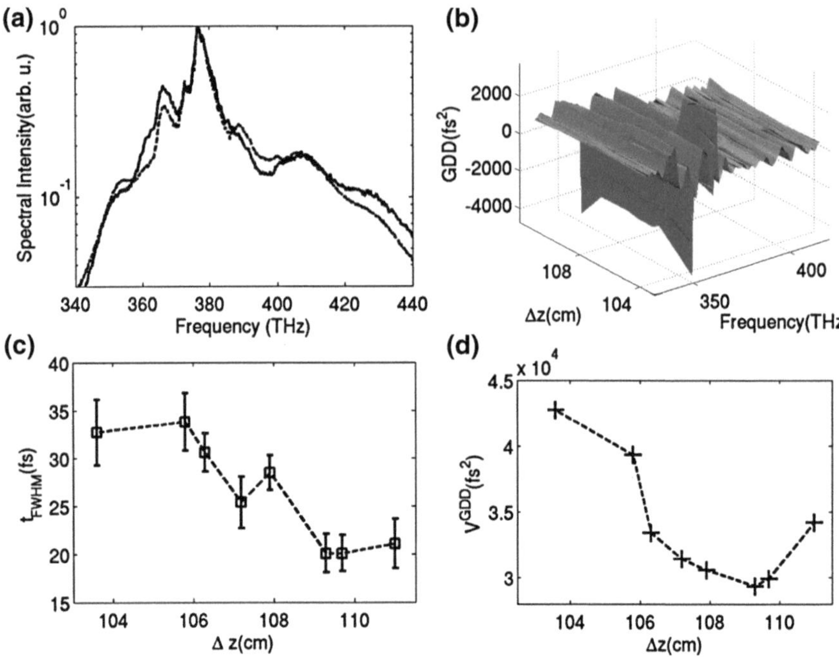

Fig. 3.19 **a** Comparison of measured spectra for small (*solid line*) and large (*dashed line*) values of Δz. Window proximity elevates the red spectral wing on the expense of the blue wing. **b** Variation of GDD versus the window position Δz. **c** Measured pulse duration versus Δz. **d** Total variation V calculated according to Eq. (3.16) in the frequency range from 340 to 425 THz. Reprinted from C. Brée et al. [81]. Copyright © 2008 American Physical Society

of the self-compression efficiency within the argon cell, and corroborate further the non-negligible influence of the cell windows.

3.4.3 Experiment 2: Windowless Measurement

In the following, the impact of the exit window is analyzed in more detail, especially the theoretically predicted dramatic temporal stretching of the pulse due to the interplay of Kerr self-focusing and GDD. To this purpose, the gas cell is positioned at $z = 103.6$ cm, which renders self-compression ineffective (see previous section). In a first experiment, the evacuated gas cell is carefully filled with argon until atmospheric pressure is reached. The laser beam is then coupled into the gas cell. The energy of the pulses entering the cell via the first aperture is 1.2 mJ. The temporal pulse profiles and spectral phases are reconstructed using the SPIDER method. In a next step, the exit window is removed from the gas cell and replaced by a 50 μm polypropylene foil. The latter is then covered by a metal plate wetted by ethanol in order to prevent an implosion of the foil during the subsequent evacuation stage. Having evacuated the gas cell, it is carefully refilled with argon until atmospheric pressure is reached, allowing to remove the metal plate. The laser beam is then coupled into the gas cell again, with identical coupling conditions as for the windowed measurements, i.e. equal pulse energy and aperture diameter. Even though the foils are a factor 10 thinner than the previously used silica windows, it is due that nonlinearity and dispersion are nearly unchanged compared to the silica window.[1] Consequently, there is no significant change in the pulse shaping dynamics observed. However, the thin foil can be easily perforated at slightly higher intensities, which enables experiments in a windowless cell. As the diameter of the aperture is very small and there is no pressure difference, it takes more than ten minutes before diffusion has significantly contaminated the argon inside the cell, as can be seen both, from a change of the fluorescence color as well as from a change in the supercontinuum spectrum. The following measurements have always been taken in the first few minutes of operation of the windowless cell, therefore minimizing the influence of air contamination. Moreover, as the ablation process perforating the foil has terminated and the power was subsequently reduced to restore input beam parameters of the windowed measurements, only plasma formation in the gases could play a role. The input beam parameters were carefully adjusted as to avoid the generation of a significant plasma density at the exit of the gas cell. The results for the windowed (red line) and windowless cases (black line) are directly compared in Fig. 3.20. In Fig. 3.20a, temporal profiles are shown. In fact, the unwindowed pulse has experienced noticeable self-compression, with a FWHM duration of 20 fs, while the windowed pulse is considerably longer (38 fs), confirming that for the chosen window position, self-compression is ineffective. The measurements impressively confirm the dramatic temporal stretching of the pulse due to the interplay of Kerr self-focusing and GDD, even exceeding the stretching predicted by linear theory for 0.5 mm of propagation in silica. The measured GDD for both cases is shown in Fig. 3.20b. Again, the fluctuations on the red side exceed those on the blue side in both cases, yet with a stronger fluctuation for the windowed pulse. This may be attributed to self-phase modulation experienced by the pulse during silica propagation.

3.4.4 Comparison with Numerical Simulations

In the following, direct numerical simulations of the evolution equation describing filamentary propagation are used in order to link our experimental results to the self-restoration results of Refs. [79, 80]. A theoretical treatment of temporal self-restoration in femtosecond filamentation requires the analysis of the pulse dynamics

---

[1] Using a band gap of 3.8 eV [83] and $n = 1.5$, an estimate of the nonlinear refractive index $n_2 = 2 \times 10^{-15}$ cm^2/W for polypropylene at 800 nm may be given. This estimation is based on Ref. [84]. A group velocity dispersion $\beta_2 = 300$ fs^2/mm is indicated in Ref. [85]. These values are at least five times larger than the characteristic values for silica at the same wavelength.

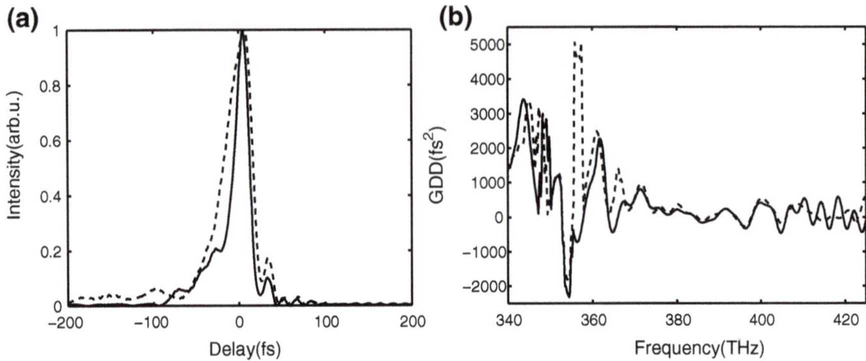

Fig. 3.20 **a** Temporal pulse profiles, and **b** GDD reconstructed from SPIDER measurements for the windowed (*dashed line*) and windowless case (*solid line*). Reprinted from C. Brée et al. [81]. Copyright © 2008 American Physical Society

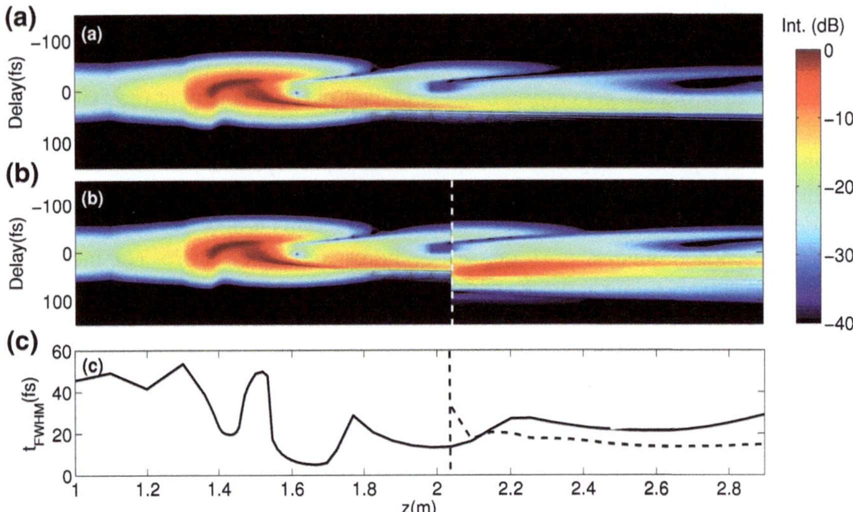

Fig. 3.21 **a** Evolution of the on-axis intensity profile for a fs-pulse propagating in argon. **b** Same, but with a silica window located at $z = 2.04\,\mathrm{m}$ ($\Delta z = 54\,\mathrm{cm}$) and subsequent propagation in air. In (**a**) and (**b**), $0\,\mathrm{dB} \cong 92\,\mathrm{TW/cm^2}$. **c** Evolution of pulse duration along z in argon (*black line*). *Dashed line* pulse duration along z for propagation in air, after traversing a 0.5 mm silica window. Reprinted from C. Brée et al. [81]. Copyright © 2008 American Physical Society

during three different propagation stages: in the first stage, the pulse propagates inside the gas cell, commonly filled with a noble gas. In the second stage, the pulse traverses the silica window, typically with a thickness of the order of 0.5 mm. Finally, before reaching the experimental setup used for pulse diagnostics, the pulse propagates in air at atmospheric pressure. The first propagation stage in the noble gas cell is appropriately described by the envelope equation Eq. (2.55) introduced in Chap. 2 of

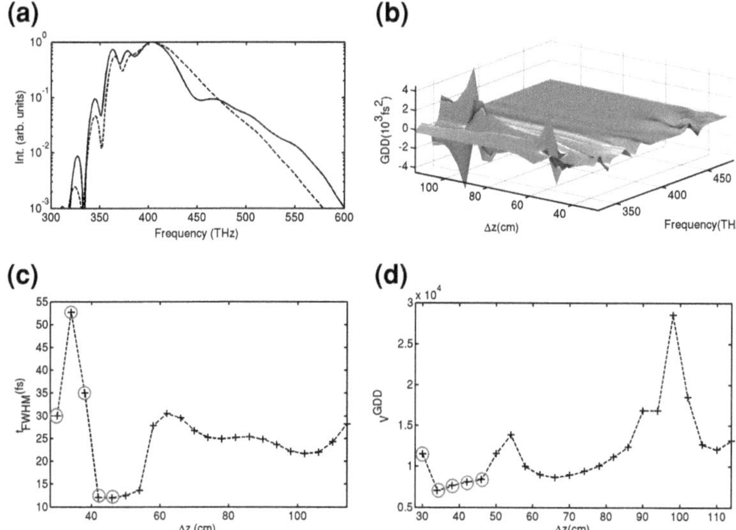

Fig. 3.22 **a** Comparison of simulated spectra for small ($\Delta z = 62$ cm, *solid line*) and large ($\Delta z = 78$ cm, *dashed line*) window-filament distance. Window proximity elevates the red spectral wing on the expense of the blue wing. **b** Variation of GDD versus window position Δz. **c** Simulated pulse duration versus Δz. **d** Total variation V^{GDD} calculated according to Eq. (3.16) in the frequency range from 340 to 425 THz. Circles in (**c**) and (**d**) indicate that in the corresponding configuration, the pulse fluence at the inner window surface exceeds 0.1 J/cm^2, leading to significant nonlinear Fresnel reflection at the argon-silica boundary. Reprinted from C. Brée et al. [81]. Copyright © 2008 American Physical Society

this thesis. However, propagation in silica and air, the former being a crystalline solid while the latter is a mixture of molecular gases, requires a refined propagation model. The main difference to the propagation Eq. (2.55) stems from the fact that the Kerr response both in silica and in air can no longer be treated as instantaneous due to the delayed response of the nuclei. Taking into account noninstantaneous contributions to the nonlinear Kerr response, the modified evolution equations are

$$\partial_z \mathcal{E} = \frac{i}{2k_0} T^{-1} \Delta_\perp \mathcal{E} + i \mathcal{D} \mathcal{E} + i \frac{\omega_0}{c} n_2 T \int \mathcal{R}(t - t') |\mathcal{E}(t')|^2 dt' \mathcal{E}$$

$$- i \frac{k_0}{2\rho_c} T^{-1} \rho(\mathcal{E}) \mathcal{E} - \frac{\sigma}{2} \rho \mathcal{E} - \frac{U_i W(I)(\rho_{nt} - \rho)}{2I} \mathcal{E}, \tag{3.17}$$

$$\partial_t \rho = W(I)(\rho_{nt} - \rho) + \frac{\sigma}{U_i} \rho I - \frac{\rho}{\tau_{\mathrm{rec}}} \tag{3.18}$$

$$\mathcal{R}(t) = (1 - f)\delta(t) + f\theta(t) \frac{1 + \omega_R^2 \tau_R^2}{\omega_R \tau_R^2} e^{-t/\tau_R} \sin(\omega_R t) \tag{3.19}$$

Here, τ_{rec} is the electron-ion recombination time, and f denotes the fractional contribution of the noninstantaneous Kerr response to the total nonlinear polarization. The function $\mathcal{R}(t)$ suitably describes the noninstataneous response in molecules or crystalline solids [86, 87]. In argon, $f = 0$, and the response kernel reduces to a δ function, describing instantaneous electronic response. On the other hand, $f = 0.15$ in silica [87] and $f = 0.5$ in air [88] account for noninstantaneous contributions to the Kerr effect, being a consequence of the delayed response of the nuclei. This is also known as the Raman effect. A pump photon excites an electron in the ground-state $|g\rangle$ to an intermediate virtual state $|i\rangle$, from which it relaxes to an excited rovibrational state $|v\rangle$ of the molecule. As the energy of the rovibrational states is higher than the ground-state energy, this is accompanied by the emission of red-shifted photon as the molecule absorbs energy from the pump field. This gives rise to the appearance of the so-called stokes line in the spectrum, redshifted w.r.t. to the pump beam. Correspondingly, the blue-shifted photon arising from the transitions $|v\rangle \rightarrow |i\rangle \rightarrow |g\rangle$ gives rise to the so-called anti-Stokes line. Equations (3.12)–(3.14) are used to analyze the dynamical behavior of a few-cycle pulse in silica sample after immediately after being self-compressed in an argon-filled gas cell. All medium parameters for argon, silica and air entering Eqs. (3.12)–(3.14) can be found tabulated in Ref. [80]. As for the initial conditions, special care was taken to numerically duplicate the parameters of the pulsed femtosecond source used in the experiment. In particular, as has been pointed out previously in literature [11, 89, 90] placing an aperture in front of the entrance window stabilizes the filament, prevents spatial break-up and may help in obtaining an increased bandwidth. Additionally, Ref. [80] has pointed out the importance of including a frequency dependence of the lens factor describing the wavefront curvature of the input pulse. Considering these two issues, an appropriate choice for the Gaussian input field is given by

$$\mathcal{E}(r, z, t) = \sqrt{\frac{2P_{\text{in}}}{\pi w_0^2}} \exp\left(-\frac{r^2}{w_0^2} - \frac{r^{16}}{r_{\text{ap}}^{16}} \right)$$

$$\times \int_{-\infty}^{\infty} d\omega \exp\left(i\frac{(\omega + \omega_0)r^2}{2cf} + i\omega t \right) \widehat{\mathcal{E}}_{\text{in}}(\omega). \qquad (3.20)$$

where $\widehat{\mathcal{E}}_{\text{in}}(\omega)$ is the Fourier transform of the assumed on-axis temporal power profile of the input pulse, $\mathcal{E}_{\text{in}}(t) = \exp\left(-t^2/t_p^2\right)$. To obtain qualitative agreement with experimental data, $t_p = 38.22\,\text{fs}$, $w_0 = 9\,\text{mm}$ and $d_{\text{ap}} = 2r_{\text{ap}} = 7\,\text{mm}$ corresponding to the diameter of the aperture are chosen as initial conditions. The focal length is given by $f = 1.5\,\text{m}$, and the input peak power P_{in} is about $84\,\text{GW}$, corresponding to 8.2 critical powers and a pulse energy of $1\,\text{mJ}$ transmitted through the aperture. The theoretical on-axis temporal profile of the initial electric field envelope, normalized to unity, is given by the Fourier transform of the spectral function $\widehat{\mathcal{E}}_{\text{in}}(\omega)$. The frequency dependent lens factor in Eq. (3.20) accounts for the fact that different frequency components diffract into different cone angles. Initially, a simulation is performed where it is assumed that the entire propagation takes place in argon. Figure 3.21a shows the numerically simulated on-axis temporal profile along propagation distance z. Here, the well-known split-isolation scheme [18, 57] is recovered, with a temporal break-up occurring around the focal range at $z = 1.5$ m and a subsequent isolation of the trailing pulse. The solid black line in Fig. 3.21c depicts the corresponding pulse duration along z. In this simulation, self-compressed pulses are obtained with a minimum duration of ≈ 20 fs, comparable to the experimentally observed scenario. Next, the latter results are compared to those obtained by accounting for the different propagation stages, i.e., argon, silica and air, as encountered under realistic experimental conditions. In air and silica, besides the instantaneous Kerr response $\sim n_2 I$, a delayed Raman term [86] contributes to the nonlinear polarization. The NEE is modified accordingly, with a relative contribution f of the delayed response to the total nonlinear polarization given by $f = 0.15$ in silica and $f = 0.5$ in air [88]. The latter values and all other medium parameters employed are tabulated in [80]. For the initial propagation stage in argon, $f = 0$, as no delayed Raman response is present in atomic gases. The complex output envelope of this simulation is then used as initial condition for the 0.5 mm propagation in silica. Finally, the output complex envelope is used as initial envelope for the last propagation stage in air.

As an example, Fig. 3.21b shows the evolution of the on-axis temporal intensity profile along z for this propagation sequence. Here the dashed white line marks the position of the exit window at $z = 2.04$ m, corresponding to $\Delta z = z - f = 54$ cm. This simulation qualitatively reproduces simulation results of [79, 80] and exhibits temporal self-restoration. This is also evidenced by the dashed line in Fig. 3.21c, exhibiting both, temporal stretching in the silica window to ≈ 33 fs from initially 14 fs, and a subsequent self-restoration of the temporal profile to 14 fs during a focusing stage. In fact, Fig. 3.21 indicates that the output window can even be beneficial for the pulse compression: For an optimum window position ($\Delta z = 54$ cm) the pulse duration for $z > 2.2$ m is even shorter than for the windowless case (see solid line in Fig. 3.21c). The window position is then further varied between $\Delta z = 34$ to 110 cm, where Δz has been chosen as the distance between the inner surface of the silica window and the focal point at $f = 1.5$ m, in analogy to the experiments. The output pulses are analyzed at $z = 2.78$ m, corresponding to the fixed position of the SPIDER setup in the experiment. Figure 3.22a shows numerical spectra for $\Delta z = 62$ cm (solid line) and $\Delta z = 78$ cm (dashed line), reproducing the experimentally observed elevation of the red spectral wing and breakdown of blue spectral wing. Figure 3.22b shows the GDD along Δz. This figure qualitatively reproduces the experimental results, showing strong GDD fluctuations in the red spectral range. In contrast, the GDD on the blue side of the spectrum has a much smaller absolute value and remains nearly constant for increasing Δz, as also evidenced experimentally. The simulated pulse duration in Fig. 3.22c strongly varies with Δz, first decreasing from 50 to 12 fs and then increasing again up to a value of ≈ 30 fs at $\Delta z = 62$ cm. Increasing Δz, the pulse duration decreases again to attain a minimum value of 22 fs at around $\Delta z = 100$ cm. For larger distances, the pulse duration increases again. Thus, the simulations reveal that the effectiveness of filamentary self-compression crucially depends upon the chosen window position. Obviously, quantitative values for the

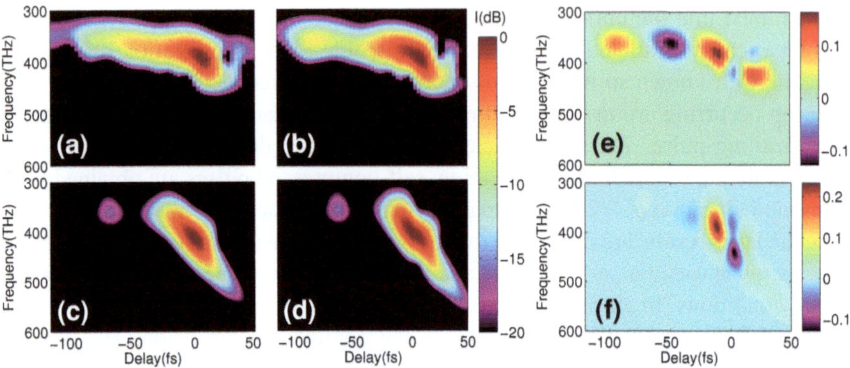

Fig. 3.23 XFROG spectrogram calculated from measured spectrum and spectral phase at **a** $\Delta z = 109.7\,$cm and **b** at $\Delta z = 103.6\,$cm. **c** Corresponding figure obtained from simulation data at $\Delta z = 78\,$cm and **d** $\Delta z = 62\,$cm. **e** Depicts numerical difference of experimental XFROG signals I_X, $\Delta I_X = I_X(\Delta z = 103.6\text{cm}) - I_X(\Delta z = 109.7\text{cm})$ shown in (**a**) and (**b**), while **f** shows the corresponding quantity for the XFROG signals shown in (**c**) and (**d**). All XFROG spectrograms are normalized to unity, $0\,$dB $\widehat{=}\,1\,$arb. u. Reprinted from C. Brée et al. [81]. Copyright © 2008 American Physical Society

measured and simulated pulse durations at comparable values for Δz (103–111 cm) disagree, in particular, the simulated pulse duration is almost constant in this Δz range. This discrepancy is to be attributed to the insufficient knowledge of the initial pulse. It is well known that pulse self-compression dynamics are very sensitive ($\sim$10 %) to input fluctuations [15]. However, as the experimentally measured pulse durations vary between 20 and 35 fs and decrease with increasing Δz (Fig. 3.19c), they are instead compared with simulated pulses in the range $60\,\text{cm} < \Delta z < 100\,\text{cm}$ exhibiting similar durations and sign of slope with respect to the window position. This latter choice is also substantiated by the behavior of $V^{\mathrm{GDD}}(\omega_1, \omega_2)$ shown in Fig. 3.22d which, at least in the interval $60\,\text{cm} < \Delta z < 80\,\text{cm}$, roughly correlates with the pulse duration. This closely reproduces the experimentally observed behavior.

Note that in Fig. 3.22c, d, for $\Delta z < 48\,$cm, the pulse fluence at the inner surface of the exit window exceeds $0.1\,\text{J/cm}^2$. According to [80, 91], this leads to significant nonlinear Fresnel reflection at the boundary, which cannot be captured by the envelope model [15, 92] employed here. Therefore, in Fig. 3.22c, d, the affected data points are highlighted by gray circles, indicating that the assumed model is strictly valid only for $\Delta z > 48\,$cm. An even deeper insight into the dynamics of temporal self-restoration is obtained by considering XFROG spectrograms [63] both from measured and simulated pulses. Figure 3.23a shows an XFROG trace calculated from a measured spectrum and SPIDER phase, corresponding to the pulse leaving the gas cell at the exit window at $\Delta z = 109.7\,$cm, a regime where self-compressed pulses with a pulse duration of 20 fs are obtained. Note that the spectrogram exhibits the well-known inverse Γ-shape discussed in previous publications [15, 57]. In contrast, positioning the exit window at $\Delta z = 103.6\,$cm, the XFROG spectrogram recon-

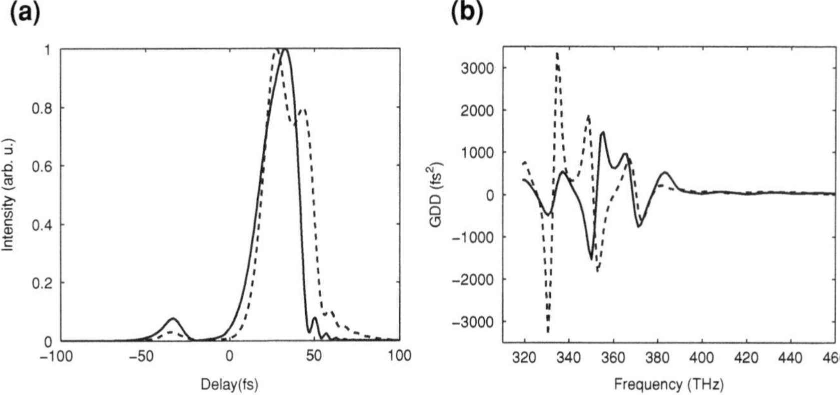

Fig. 3.24 Numerically obtained **a** on-axis temporal profile and **b** corresponding GDD for windowed (*dashed line*) and unwindowed case (*solid line*). Reprinted from C. Brée et al. [81]. Copyright © 2008 American Physical Society

structed from the measured pulse is shown in Fig. 3.23b. Here the refocusing of a self-compressed few-cycle pulse manifests itself in an increasing temporal delay of the blue spectral components, which eventually form a blue trailing subpulse. Indeed, this shift of the blue spectral components towards positive delay is observed in the experimental XFROG trace. As the effect is quite subtle, the numerical difference $\Delta I_X = I_{X,1} - I_{X,2}$ is calculated, where $I_{X,1}$ and $I_{X,2}$ denote the XFROG intensities at $\Delta z_1 = 103.6\,\mathrm{cm}$ and $\Delta z_2 = 109.7\,\mathrm{cm}$, respectively. The XFROG intensity I_X is calculated from the electric field envelope $\mathcal{E}$ according to Eq. (A.27) of Appendix C.

The visualization of ΔI_X shown in Fig. 3.23e clearly confirms the previous statement. A similar result is obtained from numerical simulations by comparing numerical XFROG traces at $\Delta z = 78\,\mathrm{cm}$ (Fig. 3.23c) and $\Delta z = 62\,\mathrm{cm}$ (Fig. 3.23d). Indeed, as revealed by the difference plot in Fig. 3.23f, the blue spectral components are shifted towards positive delays. Analyzing numerical data, it turns out that the pulse at $\Delta z = 62\,\mathrm{cm}$ undergoes a stronger refocusing event after it leaves the exit window. Therefore, in fact, according to the results of [57], its blue spectral components are expected to exhibit additional positive delay. This is evidenced both in the numerical simulations and in the experiment.

To conclude, a last numerical experiment is performed where the output pulses from windowed and unwindowed configurations are compared. For the windowless case, the simulations predict a pulse duration of $\Delta t_{\mathrm{FWHM}} = 24\,\mathrm{fs}$ at $z = 2.78\,\mathrm{m}$, whereas $\Delta t_{\mathrm{FWHM}} = 32\,\mathrm{fs}$ when a silica window is placed at $\Delta z = 62\,\mathrm{cm}$, cf. the temporal profiles shown in Fig. 3.24a. The corresponding GDD for the windowed (red line) and windowless case (black line) is shown in Fig. 3.23b. Confirming experimental observations (Fig. 3.20b), for the windowed case, a stronger fluctuation of the GDD on the red spectral wing is observed.

Even though measurements of the pulse shape or spectral phase are virtually impossible directly at location of the output window, it could be demonstrated that the window and its position are highly important in achieving a self-compressed pulse out of a filament compressor. Depending on the window position, the suggested self-healing mechanism can either be activated or inhibited. Windowless operation of the argon cell also clearly shows that the pulse coming directly out of the cell is already short, but its short temporal signature may then be spoilt by the sudden non-adiabatic change in dispersion and nonlinearity in a solid window. Our experiments indicate that the length of the highly dispersive and nonlinear material is of secondary importance, as even thin foils require self-healing in order to obtain a short pulse at the output of the cell. All these observation strongly support the theoretically predicted importance of the windows for shaping of a short pulse. These findings may explain that some authors reported problems in reproducing filament self-compression in windowed cells whereas direct self-compression in atmospheric air appeared to work right away. In summary, these observations make it manifest that future application of this versatile compression mechanism can be greatly simplified if the position of the rear window is actively adjusted for optimum shortness of the pulse.

References

1. C.V. Shank, R.L. Fork, R. Yen, R.H. Stolen, W.J. Tomlinson, Compression of femtosecond optical pulses. Appl. Phys. Lett. **40**, 761 (1982)
2. B. Schenkel, R. Paschotta, U. Keller, Pulse compression with supercontinuum generation in microstructure fibers. J. Opt. Soc. Am. B **22**, 687 (2005)
3. J. Laegsgaard, P.J. Roberts, Dispersive pulse compression in hollow-core photonic bandgap fibers. Opt. Express **16**, 9628 (2008)
4. M. Nisoli, S. DeSilvestri, O. Svelto, R. Szikops, K. Ferencz, C. Spielmann, S. Sartania, F. Krausz, Compression of high-energy laser pulses below 5 fs. Opt. Lett. **22**, 522 (1997)
5. E.B. Treacy, Optical pulse compression with diffraction gratings. IEEE J. Quantum Electron. **5**, 454 (1969)
6. S. Backus, C.G.D. III, M.M. Murnane, H.C. Kapteyn. High power ultrafast lasers. Rev. Sci. Instrum. **69**, 1207 (1997)
7. R. Szipöcs, K. Ferencz, C. Spielmann, F. Krausz, Chirped multilayer coatings for broadband dispersion control in femtosecond lasers. Opt. Lett. **19**, 201 (1994)
8. C.P. Hauri, W. Kornelis, F.W. Helbing, A. Heinrich, A. Couairon, A. Mysyrowicz, J. Biegert, U. Keller, Generation of intense, carrier-envelope phase-locked few-cycle laser pulses through filamentation. Appl. Phys. B **79**, 673 (2004)
9. A. Demircan, M. Kroh, U. Bandelow, H. Bernd, H.-G. Weber, Compression limit by third-order dispersion in the normal dispersion regime. IEEE Photonics Tech. Lett. **18**, 2353 (2006)
10. S. Kane, J. Squier, Grism-pair stretcher-compressor system for simultaneous second- and third-order dispersion compensation in chirped-pulse amplification. J. Opt. Soc. Am. B **14**, 661 (1997)
11. G. Stibenz, N. Zhavoronkov, G. Steinmeyer, Self-compression of milijoule pulses to 7.8 fs duration in a white-light filament. Opt. Lett. **31**, 274 (2006)
12. I. Koprinkov, Ionization variation of the group velocity dispersion by high-intensity optical pulses. Appl. Phys. B **79**, 359 (2004). ISSN 0946–2171, doi:10.1007/s00340-004-1553-z
13. V.F. D'Yachenko, V.S. Imshenik, Magnetohydrodynamic theory of the pinch effect in a dense high-temperature plasma (dense plasma focus). Rev. Plasma Phys. **5**, 447 (1970)
14. J.B. Taylor, Relaxation of toroidal plasma and generation of reverse magnetic fields. Phys. Rev. Lett. **33**, 1139 (1974)

15. S. Skupin, G. Stibenz, L. Berge, F. Lederer, T. Sokollik, M. Schnürer, N. Zhavoronkov, G. Steinmeyer, Self-compression by femtosecond pulse filamentation: experiments versus numerical simulations. Phys. Rev. E **74**, 056604 (2006)
16. A. Couairon, M. Franco, A. Mysyrowicz, J. Biegert, U. Keller, Pulse self-compression to the single-cycle limit by filamentation in a gas with a pressure gradient. Opt. Lett. **30**, 2657 (2005)
17. L.T. Vuong, R.B. Lopez-Martens, C.P. Hauri, A.L. Gaeta, Spectral reshaping and pulse compression via sequential filamentation in gases. Opt. Express **16**, 390 (2008)
18. C. Brée, A. Demircan, S. Skupin, L. Bergé, G. Steinmeyer, Self-pinching of pulsed laser beams during filamentary propagation. Opt. Express **17**, 16429 (2009)
19. R. Fedele, P.K. Shukla, Self-consistent interaction between the plasma wake field and the driving relativistic electron beam. Phys. Rev. A **45**, 4045 (1992)
20. N.L. Wagner, E.A. Gibson, T. Popmintchev, I.P. Christov, M.M. Murnane, H.C. Kapteyn, Self-compression of ultrashort pulses through ionization-induced spatiotemporal reshaping. Phys. Rev. Lett. **93**, 173902 (2004)
21. L. Bergé, A. Couairon, Gas-induced solitons. Phys. Rev. Lett. **86**, 1003 (2001)
22. A.M. Perelomov, V.S. Popov, M.V. Terent'ev, Ionization of atoms in an alternating electric field. Sov. Phys. JETP **23**, 924 (1966)
23. L. Bergé, A. Couairon, Nonlinear propagation of self-guided ultra-short pulses in ionized gases. Phys. Plasmas **7**, 210 (2000)
24. M.A. Darwish, On integral equations of Urysohn-Volterra type. Appl. Math. Comput. **136**, 93 (2003). ISSN 0096-3003
25. E. Babolian, F. Fattahzadeh, E.G. Raboky, A Chebyshev approximation for solving nonlinear integral equations of Hammerstein type. Appl. Math. Comput. **189**, 641 (2007)
26. C.W. Clenshaw, A.R. Curtis, A method for numerical integration on an automatic computer. Numerische Mathematik, **2**, 197 (1960). ISSN 0029-599X, doi:10.1007/BF01386223
27. S.L. Chin, Y. Chen, O. Kosareva, V.P. Kandidov, F. Théberge, What is a filament? Laser Phys. **18**, 962 (2008)
28. M. Mlejnek, E.M. Wright, J.V. Moloney, Dynamic spatial replenishment of femtosecond pulses propagating in air. Opt. Lett. **23**, 382 (1998)
29. S. Akturk, C. D'Amico, M. Franco, A. Couairon, A. Mysyrowicz, Pulse shortening, spatial mode cleaning, and intense terahertz generation by filamentation in xenon. Phys. Rev. A **76**, 063819 (2007)
30. S. Akturk, A. Couairon, M. Franco, A. Mysyrowicz, Spectrogram representation of pulse self compression by filamentation. Opt. Express **16**, 17626 (2008)
31. C. Conti, S. Trillo, P.D. Trapani, G. Valiulis, A. Piskarskas, O. Jedrkiewicz, J. Trull, Nonlinear electromagnetic X waves. Phys. Rev. Lett. **90**, 170406 (2003)
32. A. Couairon, E. Gaižauskas, D. Faccio, A. Dubietis, P.D. Trapani, Nonlinear X-wave formation by femtosecond filamentation in Kerr media. Phys. Rev. E **73**, 016608 (2006)
33. A. Zaïr, A. Guandalini, F. Schapper, M. Holler, J. Biegert, L. Gallmann, A. Couairon, M. Franco, A. Mysyrowicz, U. Keller, Spatio-temporal characterization of few-cycle pulses obtained by filamentation. Opt. Express **15**, 5394 (2007)
34. C. Brée, A. Demircan, S. Skupin, L. Bergé, G. Steinmeyer, Plasma induced pulse breaking in filamentary self compression. Laser Phys. **20**, 1107 (2010)
35. V.P. Kandidov, S.A. Shlenov, O.G. Kosareva, Filamentation of high-power femtosecond laser radiation. Quantum Electron. **39**, 205 (2009)
36. C. Brée, A. Demircan, G. Steinmeyer, Method for computing the nonlinear refractive index via Keldysh theory. IEEE J. Quantum Electron. **4**, 433 (2010)
37. R.Y. Chiao, E. Garmire, C.H. Townes, Self-trapping of optical beams. Phys. Rev. Lett. **13**, 479 (1964)
38. T.F. Coleman, Y. Li, An interior, trust region approach for nonlinear minimization subject to bounds. SIAM J. Optim. **6**, 418 (1996)
39. S. Henz, J. Herrmann, Two-dimensional spatial optical solitons in bulk Kerr media stabilized by self-induced multiphoton ionization: variational approach. Phys. Rev. E **53**, 4092 (1996)

40. E. Esarey, P. Sprangle, J. Krall, A. Ting, Self-focusing and guiding of short laser pulses in ionizing gases and plasmas. IEEE J. Quantum Electron. **33**, 1879 (1997)
41. L. Berge, S. Skupin, R. Nuter, J. Kasparian, J.P. Wolf, Ultrashort filaments of light in weakly ionized, optically transparent media. Rep. Prog. Phys. **70**, 1633 (2007)
42. S. Champeaux, L. Bergé, Postionization regimes of femtosecond laser pulses self-channeling in air. Phys. Rev. E **71**, 046604 (2005)
43. T.B. Benjamin, J.E. Feir, The disintegration of wave trains on deep water Part 1 Theory. J. Fluid Mech. **27**, 417 (1967)
44. V.I. Bespalov, V.I. Talanov, Filamentary structure of light beams in nonlinear liquids. JETP **11**, 471 (1966)
45. A. Smerzi, A. Trombettoni, P.G. Kevrekidis, A.R. Bishop, Dynamical superfluid-insulator transition in a chain of weakly coupled Bose–Einstein condensates. Phys. Rev. Lett. **89**, 170402 (2002)
46. S. Champeaux, T. Passot, P.L. Sulem, Alfvén-wave filamentation. J. Plasma Phys. **58**, 665 (1997)
47. E. Mjolhus, On the modulational instability of hydromagnetic waves parallel to the magnetic field. J. Plasma Phys. **16**, 321 (1976)
48. K. Tai, A. Hasegawa, A. Tomita, Observation of modulational instability in optical fibers. Phys. Rev. Lett. **56**, 135 (1986)
49. J.M. Dudley, G. Genty, F. Dias, B. Kibler, N. Akhmediev, Modulation instability, Akhmediev Breathers and continuous wave supercontinuum generation. Opt. Express **17**, 21497 (2009)
50. A. Demircan, U. Bandelow, Analysis of the interplay between soliton fission and modulation instability in supercontinuum generation. Appl. Phys. B **86**, 31 (2007). ISSN 0946-2171, doi:10.1007/s00340-006-2475-8
51. A. Couairon, A. Mysyrowicz, Femtosecond filamentation in transparent media. Phys. Rep. **441**, 47 (2007)
52. D.R. Solli, C. Ropers, P. Koonath, B. Jalali. Optical rogue waves. Nature **450**, 1054 (2007). ISSN 0028–0836
53. J. Kasparian, P. Béjot, J.-P. Wolf, J.M. Dudley, Optical rogue wave statistics in laser filamentation. Opt. Express **17**, 12070 (2009)
54. D.R. Solli, C. Ropers, B. Jalali, Active control of rogue waves for stimulated supercontinuum generation. Phys. Rev. Lett. **101**, 233902 (2008)
55. O. Kosareva, N. Panov, D. Uryupina, M. Kurilova, A. Mazhorova, A. Savel'ev, R. Volkov, V. Kandidov, S.L. Chin, Optimization of a femtosecond pulse self-compression region along a filament in air. Appl. Phys. B **91**, 35 (2008). ISSN 0946-2171. doi:10.1007/s00340-008-2959-9
56. P. Hauri, R.B. Lopez-Martens, C.I. Blaga, K.D. Schultz, J. Cryan, R. Chirla, P. Colosimo, G. Doumy, A.M. March, C. Roedig, E. Sistrunk, J. Tate, J. Wheeler, L.F. DiMauro, E.P. Power, Intense self-compressed, self-phase-stabilized few-cycle pulses at 2 μm from an optical filament. Opt. Lett., **32**, 868 (2007)
57. C. Brée, J. Bethge, S. Skupin, L. Bergé, A. Demircan, G. Steinmeyer, Cascaded self-compression of femtosecond pulses in filaments. New J. Phys. **12**, 093046 (2010)
58. A. Dalgarno, A.E. Kingston, The refractive indices and Verdet constants of the intert gases. Proc. Roy. Soc. A **259**, 424 (1960)
59. H.J. Lehmeier, W. Leupacher, A. Penzkofer, Nonresonant third order hyperpolarizability of rare gases and N2 determined by third order harmonic generation. Opt. Commun. **56**, 67 (1985)
60. D. Faccio, A. Averchi, A. Lotti, P.D. Trapani, A. Couairon, D. Papazoglou, S. Tzortzakis, Ultrashort laser pulse filamentation fromspontaneous X wave formation in air. Opt. Express **16**, 1565 (2008)
61. C. Brée, A. Demircan, G. Steinmeyer, Modulation instability in filamentary self-compression. Laser Phys. **21**, 1313 (2011). ISSN 1054–660X. doi:10.1134/S1054660X11130044
62. S. Eisenmann, A. Pukhov, A. Zigler, Fine structure of a Laser-plasma filament in air. Phys. Rev. Lett. **98**, 155002 (2007)

63. S. Linden, H. Giessen, J. Kuhl, XFROG—A new method for amplitude and phase characterization of weak ultrashort pulses. physica status solidi (b) **206**, 119 (1998). ISSN 1521-3951
64. J. Dudley, X. Gu, L. Xu, M. Kimmel, E. Zeek, P. O'Shea, R. Trebino, S. Coen, R. Windeler, Cross-correlation frequency resolved optical gating analysis of broadband continuum generation in photonic crystal fiber: simulations and experiments. Opt. Express **10**, 1215 (2002)
65. A.L. Gaeta, Catastrophic collapse of ultrashort pulses. Phys. Rev. Lett. **84**, 3582 (2000)
66. M.A. Porras, A. Parola, D. Faccio, A. Couairon, P.D. Trapani, Light-filament dynamics and the spatiotemporal instability of the Townes profile. Phys. Rev. A **76**, 011803(R) (2007)
67. L. Bergé, J.J. Rasmussen, Multisplitting and collapse of self-focusing anisotropic beams in normal/anomalous dispersive media. Phys. Plas. **3**, 824 (1996)
68. L. Berge, K. Germaschewski, R. Grauer, J.J. Rasmussen, Hyperbolic shockwaves of the optical self-focusing with normal group-velocity dispersion. Phys. Rev. Lett. **89**, 153902 (2002)
69. G.P. Agrawal, *Nonlinear Fiber Optics*, 3rd edn. (Academic Press, San Diego, 2001)
70. D. Faccio, M.A. Porras, A. Dubietis, F. Bragheri, A. Couairon, P.D. Trapani, Conical emission, pulse splitting, and X-wave parametric amplification in nonlinear dynamics of ultrashort light pulses. Phys. Rev. Lett. **96**, 193901 (2006)
71. C. Iaconis, I.A. Walmsley, Spectral phase interferometry for direct electric-field reconstruction of ultrashort optical pulses. Opt. Lett. **23**, 792 (1998)
72. C. Iaconis, I.A. Walmsley, Self-referencing spectral interferometry for measuring ultrashort optical pulses. IEEE J. Quantum Electron. **35**, 501 (1999)
73. L. Gallmann, D.H. Sutter, N. Matuschek, G. Steinmeyer, U. Keller, C. Iaconis, I.A. Walmsley, Characterization of sub-6-fs optical pulses with spectral phase interferometry for direct electric-field reconstruction. Opt. Lett. **24**, 1314 (1999)
74. B.A. Malomed, D. Mihalache, F. Wise, L. Torner, Spatiotemporal optical solitons. J. Opt. B **7**, R53 (2005)
75. I.G. Koprinkov, A. Suda, P. Wang, K. Midorikawa, Self-compression of high-intensity femtosecond optical pulses and spatiotemporal soliton generation. Phys. Rev. Lett. **84**, 3847 (2000)
76. A.L. Gaeta, F. Wise, Comment on self-compression of high-intensity femtosecond optical pulses and spatiotemporal soliton generation. Phys. Rev. Lett. **87**, 229401 (2001)
77. S. Skupin, L. Bergé, U. Peschel, F. Lederer, Interaction of femtosecond light filaments with obscurants in aerosols. Phys. Rev. Lett. **93**, 023901 (2004)
78. C. Sulem, P.-L. Sulem, *The Nonlinear Schrödinger Equation: Self-Focusing and Wave Collapse*, Applied Mathematical Sciences, (Springer, New York, 1999)
79. L. Berge, S. Skupin, G. Steinmeyer, Temporal self-restoration of compressed optical filaments. Phys. Rev. Lett. **101**, 213901 (2008). doi:10.1103/PhysRevLett.101.213901
80. L. Bergé, S. Skupin, G. Steinmeyer, Self-recompression of laser filaments exiting a gas cell. Phys. Rev. A **79**, 033838 (2009)
81. C. Brée, A. Demircan, J. Bethge, E.T.J. Nibbering, S. Skupin, L. Bergé, G. Steinmeyer, Filamentary pulse self-compression: the impact of the cell windows. Phys. Rev. A **83**, 043803 (2011). doi:10.1103/PhysRevA.83.043803
82. G. Stibenz, G. Steinmeyer, Optimizing spectral phase interferometry for direct electric-field reconstruction. Rev. Sci. Instrum. **77**, 073105 (2006)
83. S.A.Y. Al-Ismail, C.A. Hogarth, Optical absorption of metal-loaded polymer films prepared by vacuum evaporation. J. Mater. Sci. Lett. **7**, 135 (1988). ISSN 0261–8028, doi:10.1007/BF01730595
84. M. Sheik-Bahae, D.C. Hutchings, D.J. Hagan, E.W. van Stryland, Dispersion of bound electronic nonlinear refraction in solids. IEEE J. Quantum Electron. **27**, 1296 (1991)
85. T. Yovcheva, T. Babeva, K. Nikolova, G. Mekishev, Refractive index of corona-treated polypropylene films. J. Opt. A Pure Appl. Opt. **10**, 055008 (2008)
86. P. Sprangle, J.R. Pe nano, B. Hafizi, Propagation of intense short laser pulses in the atmosphere. Phys. Rev. E **66**, 046418 (2002)
87. A.A. Zozulya, S.A. Diddams, A.G.V. Engen, T.S. Clement, Propagation ynamics of intense femtosecond pulses: multiple splittings, coalescence, and continuum generation. Phys. Rev. Lett. **82**, 1430 (1999)

88. E.T.J. Nibbering, G. Grillon, M.A. Franco, B.S. Prade, A. Mysyrowicz, Determination of the inertial contribution to the nonlinear refractive index of air, N2, and O2 by use of unfocused high-intensity femtosecond laser pulses. J. Opt. Soc. Am. B **14**, 650 (1997)
89. J.-F. Daigle, O. Kosareva, N. Panov, M. Bégin, F. Lessard, C. Marceau, Y. Kamali, G. Roy, V. Kandidov, S.L. Chin, A simple method to significantly increase filaments' length and ionization density. Appl. Phys. B **94**, 249 (2009)
90. J. Bethge, C. Brée, H. Redlin, G. Stibenz, P. Staudt, G. Steinmeyer, A. Demircan, S. Düsterer, Self-compression of 120 Å fs pulses in a white-light filament. J. Opt. **13**, 055203 (2011)
91. J.R. Peñano, P. Sprangle, B. Hafizi, W. Manheimer, A. Zigler, Transmission of intense femtosecond laser pulses into dielectrics. Phys. Rev. E **72**, 036412 (2005)
92. T. Brabec, F. Krausz, Nonlinear optical pulse propagation in the single-cycle regime. Phys. Rev. Lett. **78**, 3282 (1997)

Chapter 4
Saturation and Inversion of the All-Optical Kerr Effect

Recently, it has been shown both, experimentally and theoretically [1–3] that for intensities relevant in femtosecond filamentation, the Kerr refractive index of major air components exhibits a saturation behavior and changes its sign from a focusing to a defocusing nonlinearity, as is shown in Fig. 4.1 taken from [2]. In Ref. [4], the implications of this surprising behavior on femtosecond filamentation have been analyzed using numerical simulations. As the unexpected saturation behavior clearly cannot be theoretically modeled by truncating the power series for the Kerr refractive index after the n_2 term, this underlines the urgent need for a theoretical determination of the higher order nonlinearities. The observed saturation behavior is completely contrary to the present model of filamentary propagation, as up to now it has widely been accepted that the refractive index change induced by free electrons provides the dominant saturation mechanism counteracting Kerr self-focusing. Therefore, the recent development has led the authors of Ref. [5] to postulate the possibility of a "paradigm shift", and they propose an experiment designed to clarify the role of the higher order nonlinearities by measuring the efficiency of fifth harmonic generation in the medium under consideration. Therefore, the recent indications of a dominant role of higher-order nonlinearities clearly require further investigation.

In the current chapter, a theoretical estimate on the expansion coefficients n_{2k} of the intensity dependent refractive index

$$n(I) = \sum_{k \geq 0} n_{2k} I^k \tag{4.1}$$

for different noble gases is provided. Based on the theoretical investigations of Ref. [6], Kramers-Kronig (KK) theory is used to provide theoretical estimates of arbitrary higher order nonlinearities. As conjectured in Ref. [3], the saturation behavior of the Kerr refractive index is closely related to ionization of the noble gas atoms by the intense laser field. Using a recently developed model [7] for ionization of atoms in strong alternating electric fields, cross sections σ_K for the ionization of the atoms by simultaneous absorption of K photons are calculated. This makes it pos-

C. Brée, *Nonlinear Optics in the Filamentation Regime*, Springer Theses,   79
DOI: 10.1007/978-3-642-30930-4_4, © Springer-Verlag Berlin Heidelberg 2012

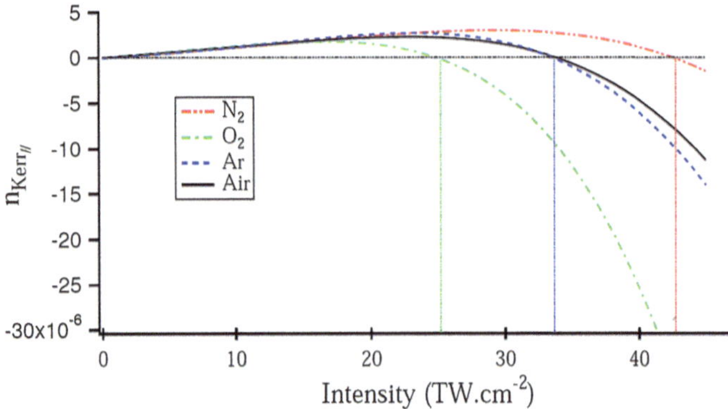

Fig. 4.1 Intensity dependent refractive index for major air components versus intensity at standard conditions. The figure is taken from Ref. [2]. Reprinted from V. Loriot et al. [2]. Copyright © 2010 Optical Society of America

sible to relate the calculated multiphoton absorption spectra, via KK theory, to the higher order nonlinear coefficients $n_{2(K-1)}$. In Sect. 4.1, KK theory in linear optics and its generalization to nonlinear optical susceptibilities is briefly reviewed. This method has first been used to calculate the nonlinear refractive index n_2 from the cross-section for two-photon absorption (TPA) between the valence and conduction band of certain semiconductors [8, 9]. In fact, the remarkably accurate results of the approach obtained in Ref. [8] provided a strong motivation to transfer the principles and results of the work of Sheik-Bahae et al. to the nonlinear optical response of noble gas atoms, as this is highly relevant for the modeling of femtosecond filamentation.

As knowledge of multiphoton cross-sections is a prerequisite for calculating the nonlinear refractive index using the KK transform, in Sect. 4.2 the prevalent theories describing the ionization of atoms in intense laser-fields are briefly discussed. Using a recent modification of Popov-Perelomov-Terent'ev (PPT) theory [7, 10], cross-sections for multiphoton absorption of atomic gases are derived. Extending the results of Ref. [11], in Sect. 4.3, KK theory is used to calculate the nonlinear refractive index n_2 from TPA cross-sections in helium, neon, argon, krypton and xenon. Comparing the results on the dispersion of n_2 with values established in the literature, excellent agreement is found, especially in the long-wavelength region. In Sect. 4.4, KK-theory is used to obtain numerical values of the higher-order nonlinearities n_{2k}, which allows calculating the intensity dependent refractive index change induced by the Kerr effect, $\Delta n(I) \equiv n(I) - n_0$ up to arbitrary order in I, observing the recently predicted saturation behavior. A brief discussion is given, and it is pointed out that the obtained results, together with independently obtained ones, may have paradigm changing consequences for the theoretical modeling of femtosecond filamentation.

4.1 Kramers-Kronig Relations in Linear and Nonlinear Optics

Any theoretical description of physical reality is subject to the requirement of causality. In the framework of Newtonian mechanics, this simply states that the state of any given physical system at time t is only affected by events occurring at $t' < t$. In linear optics, the requirement of causality has led to the formulation of the KK relations [12, 13] between the real and imaginary part of the linear susceptibility. The time domain analog of Eq. (2.7) for the linear polarization is given by

$$P(\vec{r}, t) = \epsilon_0 \int_{-\infty}^{\infty} d\tau R^{(1)}(\tau) E(\vec{r}, t - \tau), \tag{4.2}$$

where the optical response is governed by a response kernel $R^{(1)}(t)$ which is related to the linear susceptibility $\chi^{(1)}(\omega)$ via a Fourier transform. The requirement of causality manifests itself in the identity

$$R(t) = R(t)\Theta(t), \tag{4.3}$$

where $\Theta(t)$ denotes the Heaviside step function defined by $\Theta(t) = 1$ for $t > 0$ and $\Theta(t) = 0$ for $t < 0$. Thus, the Eqs. (4.2) and (4.3) simply state that only field configurations from the past, $E(t')$ with $t' < t$, can affect the linear optical response. Then, the frequency domain analogue of Eq. (4.3) provides the KK relation, which reads

$$\chi(\omega) = \frac{1}{i\pi} \mathcal{P} \int_{-\infty}^{\infty} \frac{\chi(\Omega)}{\Omega - \omega} d\Omega, \tag{4.4}$$

where $\mathcal{P}$ denotes Cauchy's principal value. The more familiar form of the KK relations relates dispersion $n(\omega)$ and absorption coefficients $\alpha(\omega)$ according to

$$n(\omega) - 1 = \frac{c}{\pi} \mathcal{P} \int_{0}^{\infty} \frac{\alpha(\Omega)}{\Omega^2 - \omega^2} d\Omega. \tag{4.5}$$

This relation is completely equivalent to Eq. (4.4), as linear dispersion $n(\omega)$ and absorption $\alpha(\omega)$ are related to the real and the imaginary part of χ according to Eq. (2.44). In nonlinear optics, the nonlinear polarization P_{NL} may be represented as a power series in the electric field components according to $P_{\mathrm{NL}} = P^{(3)} + P^{(5)} + ...$, where the n-th order contribution is given by Eq. (2.20). In order to respect causality, the response function $R^{(n)}$ must satisfy

$$R^{(n)}(\tau_1, \tau_2, ..., \tau_n) = R^{(n)}(\tau_1, \tau_2, ..., \tau_n)\Theta(\tau_i) \tag{4.6}$$

for any $i = 1, 2, ..., n$. Again, taking the Fourier transform of this relation with respect to all time arguments, it is straightforward to see that the n-th order nonlinear susceptibility satisfies the KK-type relation

$$\chi^{(n)}(-\omega_\sigma; \omega_1, \omega_2, ..., \omega_i, ..., \omega_n) = \frac{1}{i\pi}\mathcal{P}\int_{-\infty}^{\infty} \frac{\chi^{(n)}(-\omega_\sigma; \omega_1, \omega_2, ..., \Omega, ..., \omega_n)}{\Omega - \omega_i}d\Omega.$$

(4.7)

As in linear optics, for the case of nonlinear refraction of a probe beam induced by a strong pump beam this may be recast into an equation relating the refractive index change Δn to the change in absorption $\Delta\alpha$ according to

$$\Delta n(\omega; \xi) = \frac{c}{\pi}\mathcal{P}\int_{0}^{\infty} \frac{\Delta\alpha(\Omega; \xi)}{\Omega^2 - \omega^2}d\Omega.$$

(4.8)

Here, according to Ref. [14], the variable ξ represents the source of the absorption and refractive index changes. For example, consider a probe beam at frequency ω_1 and a strong pump beam at ω_2 which induces a change of the refractive index seen by the probe beam. Then, $\xi = \omega_2$, $\Delta n(\omega_1, \omega_2) = \frac{3}{4n_0^2\epsilon_0 c}\text{Re}\chi^{(3)}(-\omega_1; \omega_1, \omega_2, -\omega_2)I$ and $\Delta\alpha(\omega_1, \omega_2) \propto \text{Im}\chi^{(3)}(-\omega_1; \omega_1, \omega_2, -\omega_2)I$ describes nondegenerate TPA, cf. Eqs. (2.50), (2.51). In the following, this relation will be used to calculate n_2 and higher-order nonlinearities from known multiphoton cross-sections in helium, neon, argon, krypton and xenon.

4.2 Ionization of Atoms in Intense Laser Fields

Since the development of laser sources delivering pulse energies of several μJ at pulse durations of only a few cycles of the optical carrier field, optical intensities can be achieved which suffice to ionize a considerable fraction of the atoms of the propagation medium. Considering femtosecond filamentation, where typically air or noble gases are used in experiments, according to the classical theoretical model of filamentation, the intensity in the vicinity of the nonlinear focus is of the order of the clamping intensity. For argon at atmospheric pressure, this intensity corresponds to ≈ 100 TW/cm^2. However, considering the ionization potential of argon, photon energies of $\hbar\omega \geq U_i = 15.7596$ eV are necessary for photo-ionization of the gas, corresponding to a laser wavelength of $\lambda \approx 80$ nm in the XUV regime, whereas femtosecond filamentation is normally considered in the visible or near IR regime. To resolve this apparent contradiction, it can be shown that for sufficiently high intensities, higher-order terms in the perturbative expansion of the ionization cross-section start to contribute. Using lowest-order perturbation theory (LOPT) [15, 16], it is found that the K-th-order contribution to the ionization rate is given by

$w_K = \sigma_K I^K$, with the cross-section

$$\sigma_K \propto \left| \sum_{a_{K-1}} \cdots \sum_{a_1} \frac{\langle f | \vec{\mu} \cdot \vec{\epsilon} | a_{K-1} \rangle \cdots \langle a_1 | \vec{\mu} \cdot \vec{\epsilon} | g \rangle}{[E_{a_{K-1}} - E_g - (K-1)\hbar\omega] \cdots (E_{a_1} - E_g - \hbar\omega)} \right|^2. \qquad (4.9)$$

In this expression, $|f\rangle$ and $|g\rangle$ denote the ground- and final electronic state, while $|a_k\rangle$ denotes a complete set of atomic states. The corresponding energy levels are given by E_g, E_{a_l}, and $\vec{\mu} \cdot \vec{\epsilon}$ denotes the projection of the electronic dipole operator onto the polarization direction $\vec{\epsilon}$ of the incident electric field. The rate w_K describes multiphoton ionization (MPI) of an atom by simultaneous absorption of K photons, such that the total energy of the absorbed photon exceeds the ionization potential, i.e., $K\hbar\omega \geq U_i$. MPI can adequately be described using perturbation theory. For example, LOPT has successfully been applied to obtain multiphoton cross sections for evaporated metal atoms [17]. Combining LOPT with an *ab initio* method for the calculation of two-electron wavefunctions, MPI cross-sections of molecular hydrogen were obtained [18].

The precise knowledge of MPI rates w_K is crucial in order to obtain higher-order Kerr coefficients by using the generalized KK relations Eq. (4.8). Indeed, K-photon ionization gives rise to a nonlinear absorption change according to a power law intensity dependence $\Delta\alpha = \beta_K I^{K-1}$ (cf. Sect. 4.4). Given the validity of the perturbative description of the nonlinear polarization Eq. (2.19) resp. that of the intensity dependent refractive index $n(I)$ Eq. (2.49), it follows that $\Delta\alpha$ can be related to the $K-1$-th order contribution $\Delta n = n_{2(K-1)} I^{K-1}$ via the generalized KK relation Eq. (4.8). It is, however, important to point out that there exist regimes where the perturbative multiphoton description of ionization of atoms in strong laser fields breaks down, and so called tunneling ionization becomes relevant. To identify the different regimes, in his pioneering work, Keldysh [19] analyzed strong field ionization of hydrogen like atoms and introduced the so-called Keldysh parameter

$$\gamma = \frac{\omega\sqrt{2\hbar\omega_p m_e}}{E_0 q_e}. \qquad (4.10)$$

Here, E_0 denotes the amplitude of the electric field, ω the frequency of the applied laser field, m_e and q_e denote electron mass and charge, respectively, and $\hbar\omega_p = U_i$ is the ionization potential of the gas species. The tunneling picture depends on the formation of a Coulomb barrier due to the superposition of the optical potential with the atomic Coulomb potential. This can only be satisfied when the strength of the ponderomotive potential[1] $U_p = q_e^2 E_0^2 / 4m_e \omega^2$ exceeds that of the ionization potential U_i. By noting that the Keldysh parameter Eq. (4.10) may be written as $\gamma = \sqrt{U_i / 2U_p}$, it follows that the tunneling regime corresponds to the limit $\gamma \ll 1$. In this limit, the applied ac electric field deforms the Coulomb potential as shown in Fig. 4.2, allowing the bound electron to tunnel through the resulting barrier. In fact,

---

[1] The ponderomotive potential corresponds to the cycle averaged quiver energy of an electron in an external electromagnetic field.

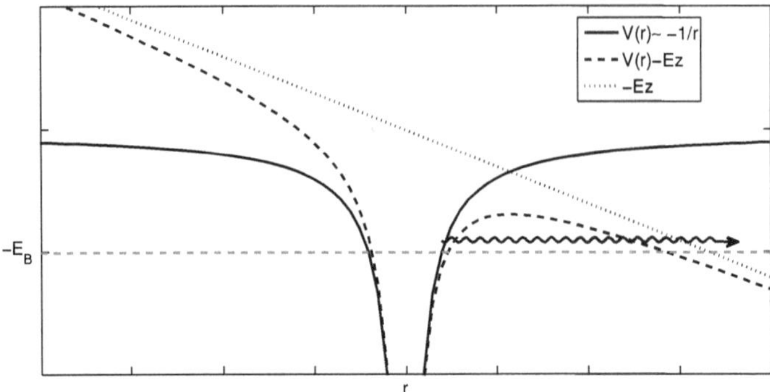

Fig. 4.2 a Tunnel ionization. *Black solid line* represents the undistorted Coulomb potential. The *black dotted line* corresponds to the linear potential Ez due to the presence of the electric field E, and the *dashed black line* represents the superposition of both atomic and laser potential. The *gray dashed line* corresponds to the binding energy $-E_B$ of the electron

following Keldysh's work, a variety of models based on and extending Keldysh theory were developed. The prevalent models employed in today's strong field physics are the KFR (Keldysh [19], Faisal [20] , Reiss [21]) theories. While Keldysh theory is a tunneling model assuming $\gamma \ll 1$, the Faisal model is a high-frequency approximation requiring $\omega \gg \omega_p$ and $U_p \gg U_i$. The model of Ref. [21] requires only $U_p \gg U_i$, which is consequently referred to as the **strong field approximation** (SFA). SFA uses a rigorous S-matrix formalism to justify the necessary approximations and provides photo-electron spectra which accurately agree with measured spectra. While SFA provides excellent photo-electron spectra, the ionization model provided by Perelomov, Popov and Terent'ev provides more accurate results when only total ionization rates are required. In fact, in [22] it has been shown that the ionization rate of PPT accurately fits experimental data, whereas SFA underestimates the data by 2–3 orders of magnitude. In the following, the ionization rate provided by PPT will be used. A theoretical treatment of the ionization of atoms in intense laser fields and a recent modification of PPT-theory is presented, giving accurate results both in the tunneling regime ($\gamma \ll 1$) and especially in the multiphoton regime ($\gamma \gg 1$). In fact, the perturbative limit $w(I) = \sigma_K I^K$ provides absorption cross-sections $\sigma_K(\omega)$ which can be used to calculate the expansion coefficients of the power series for $n(I)$ and their dispersion with frequency.

4.2.1 Keldysh Theory and Its Generalizations

In this subsection, the theoretical foundations of PPT theory are briefly presented. Only the basic physical assumptions and approximations are discussed here, as it is not the aim of this subsection to rederive the ionization rate of [10] in detail. For the current discussion, all quantities are given in atomic units. However, the final result, i.e., the ionization rate and the multiphoton cross sections used in the subsequent section, are presented in SI units for convenience. In atomic units, the problem of ionization of atoms in an intense laser field is described by the Schrödinger equation $i\partial_t\psi = H\psi$, with a Hamiltonian given by $H = -\vec{p}^2/2 + V(\vec{r}) - \vec{E}\cdot\vec{r}$, where $V(\vec{r})$ is the binding potential of the atom, $\vec{p} = (p_x, p_y, p_y)$ and $\vec{r} = (x, y, z)$ are momentum and space coordinates, respectively, and $\vec{E}$ is the electric field. It is possible to show that the Schrödinger equation is completely equivalent to the following integral equation

$$\psi(\vec{r}, t) = \int d^3\vec{r}'G(\vec{r}, t; \vec{r}', t')\psi(\vec{r}', t_0) + \int\limits_{t_0}^{t} dt \int\limits_{\mathbb{R}^3} d^3\vec{r}'G(\vec{r}, t; \vec{r}', t')V(\vec{r}')\psi(\vec{r}', t)$$

$$(4.11)$$

where $G(\vec{r}, t; \vec{r}', t')$ denotes the Greens function of an electron in the external potential $-\vec{E}\cdot\vec{r}$. With ψ_α the discrete or continuous eigenfunction and E_α the corresponding eigenenergies, the Greens function may, quite generally, be written as $G(\vec{r}, t; \vec{r}', t') = \sum \psi_\alpha(\vec{r})\psi_\alpha^*(\vec{r}') \times \exp\left(-iE_\alpha(t - t')\right)$. For the present problem, it is given as

$$G(\vec{r}, t; \vec{r}', t') = \theta(t - t')\int d^3\vec{p}\,\Psi_V(\vec{r}, t)\Psi_V^*(\vec{r}', t')$$

$$(4.12)$$

The continuum wavefunction Ψ_V denotes the non-relativistic Volkov state [23] and solves the Schrödinger equation of a charged particle in an electromagnetic field. It is given by

$$\Psi_V(\vec{r}, t) = \frac{1}{(2\pi)^{3/2}}\exp\left(i\left[\vec{\pi}(t)\cdot\vec{r} - \frac{1}{2}\int\limits_{-\infty}^{t}\vec{\pi}^2(\tau)d\tau\right]\right) \qquad (4.13)$$

The generalized momentum $\vec{\pi}$ couples the vector potential $\vec{A}$ to the electronic wavefunction. These quantities are related to the kinematic momentum $\vec{p}$ and the electric field $\vec{E}$, respectively, according to

$$\vec{\pi}(t) = \vec{p} - \vec{A}(t), \quad \vec{A}(t) = -\int\limits_{-\infty}^{t}\vec{E}(t')dt'. \qquad (4.14)$$

The spatially uniform electric field is chosen as

$$\vec{E}(t) = \vec{E}_0\cos\omega t, \qquad (4.15)$$

where it is assumed that the field is linearly polarized along the x-axis, $\vec{E}_0 = (E_0, 0, 0)$. The field is turned on adiabatically at $t_0 \to -\infty$. Note that although the derivation of PPT assumes a monochromatic wave, their results are commonly generalized for electric fields of the form $\vec{E}(t) = \vec{E}_0(t)\cos\omega t$ with a time dependent envelope $E_0(t)$. With the decompositions (2.27) and (2.28), one then finds that the amplitude $E_0(t)$ equals the modulus of the complex envelope, $E_0(t) = |\mathcal{A}(t)|$. Then, the Keldysh parameter may be written

$$\gamma = \frac{\omega\sqrt{2\hbar\omega_p m_e}}{|\mathcal{A}(t)|q_e}. \tag{4.16}$$

As only cycle-averaged ionization rates are provided by PPT, this generalization which involves a time-dependent Keldysh parameter, is meaningful for envelopes $E_0(t)$ varying slowly with respect to the optical cycle. This issue was dealt with in detail in Ref. [24], providing an ionization rate that follows the instantaneous phase of the electric field.

Note that here, as in the original work of Keldysh [19], the so called length gauge is employed, where the electromagnetic interaction contributes with a term $\sim -\vec{r}\cdot\vec{E}$ to the Hamiltonian. In velocity gauge, the same interaction is modeled by coupling the vector potential to the electron wavefunction according to the prescription $\vec{p} \to \vec{p} - \vec{A}$, i.e., in this gauge, the Hamiltonian is given by $(\vec{p} - \vec{A})^2/2$. Clearly, any measurable quantity within a theoretical model of physical reality should be gauge-invariant. However, a perturbative treatment of the interaction of electrons with strong laser fields may break gauge invariance, and the initial choice of gauge strongly affects the numerical outcomes of the corresponding theory. In SFA, the impact of choosing a specific gauge on photo-electron spectra and ionization rates has been extensively discussed in [25].

Equation (4.11) may be further simplified by noting that the first term on the r.h.s. describes the motion of an otherwise free electron in a spatially uniform alternating electric field. Thus, this term gives no contribution to the ionization current. The main approximation used in Ref. [10] is then to replace the exact wave function $\psi(\vec{r}, t)$ by the wave function $\phi_{\ell m}(\vec{r})\exp(i\kappa^2 t/2)$ of the unperturbed atomic bound state, where ℓ, m are angular and magnetic quantum numbers, respectively. This approximation is justified if the electric field strength $\vec{E}$ is smaller than the internal atomic field strength, i.e., $E_0 \ll \Theta$, where, in atomic units, $\Theta = \kappa^3$. In SI units, this quantity amounts to $\Theta = \frac{4}{q}\omega_p^{3/2}\sqrt{2m_e\hbar}$. This condition ensures that near the atomic nucleus, i.e., for $\kappa r < 1$, the exact electronic wavefunction nearly coincides with its unperturbed counterpart $\phi_{\ell m}(\vec{r})\exp(i\kappa^2 t/2)$. For $\kappa r \gg 1$, this substitution is only justified when the atomic potential $V(\vec{r})$ vanishes faster than $1/r$. Then, from regions $\kappa r > 1$ where ψ differs from the unperturbed wavefunction, a negligible contribution to the integral in Eq. (4.11) results. These approximations are equivalent to assuming a Volkov state [23] as the final state of the ionized electron.[2]

---

[2] The limitations of these approximations can be partially overcome by introducing dressed final or initial electronic states. Dressed atomic states were used in Refs. [26, 27] as initial states to account, e.g., for the shift of atomic energy levels due to the applied AC electric field. These field-dependent bound states are derived on account of Floquet's theorem applied to the Schrödinger equation with a time periodic perturbation due to the irradiated laser field. To account for the long-range Coulomb potential, dressed continuum states have been derived in Ref. [28], as discussed below.

Finally, $\psi(\vec{r}, t)$ is replaced by $\phi_{\ell m}(\vec{r}) \exp(i\kappa^2 t/2)$. As the latter function solves the stationary Schrödinger equation, on the r.h.s of Eq. (4.11) one may replace $V(\vec{r})\phi_{\ell m}$ by $(\vec{\nabla}^2 - \kappa^2)\phi_{\ell m}(\vec{r})/2$ to obtain the following estimate for the perturbed wave function,

$$\psi(\vec{r}, t) = \int_{-\infty}^{t} dt' \int d^3\vec{r}' G(r, r', t, t') \frac{1}{2}(\vec{\nabla}'^2 - \kappa^2)\phi_{\ell m}(\vec{r}') \exp(i\kappa^2 t'/2), \quad (4.17)$$

where $\vec{\nabla}' = (\partial/\partial x', \partial/\partial y', \partial/\partial z')$. Assuming that the incident electric field is linearly polarized along the x-axis, using Eqs. (4.12) and (4.17) one calculates the flux of electrons through the plane perpendicular to the polarization axis according to

$$J(x, t) = \int dy \int dz\, j_x(\vec{r}, t), \quad (4.18)$$

where the x-component j_x of the probability current $\vec{j}$ is given by

$$j_x(\vec{r}, t) = \frac{i}{2}\left(\psi(\vec{r}, t)\frac{\partial}{\partial x}\psi^*(\vec{r}, t) - \psi^*(\vec{r}, t)\frac{\partial}{\partial x}\psi(\vec{r}, t)\right). \quad (4.19)$$

In order to calculate the probability current, the approximate expression Eq. (4.17) for the wavefunction is employed. Finally, the ionization rate may be obtained from taking the limit

$$w(E, \omega) = 2 \lim_{x \to \infty} \overline{J(x, t)}. \quad (4.20)$$

where $\overline{J(x, t)}$ represents the cycle averaged ionization current. Interestingly, the ionization rate may be decomposed into a sum over probabilities of n-photon processes according to

$$w(E, \omega) = \sum_{n \geq \nu}^{\infty} w_n(E, \omega), \quad \text{with } \nu = \frac{\omega_p}{\omega}\left(1 + \frac{1}{2\gamma^2}\right). \quad (4.21)$$

The summation starts with the smallest natural number n_0 greater than ν, accounting for the fact that at least n_0 photons are necessary to provide both, the binding energy ω_p necessary for ionizing the atom, and the ponderomotive energy $U_p = \kappa^2/4\gamma^2$. The quantity $w_n(E, \omega)$ is given by

$$w_n(E, \omega) = 2\pi \int d^3\vec{p}\,\delta\left(\frac{1}{2}\left[\vec{p}^2 + \kappa^2 + \frac{\kappa^2}{2\gamma^2}\right] - n\omega\right)|F_n(\vec{p})|^2. \quad (4.22)$$

Again, the δ-function in the integrand enforces conservation of energy, i.e., the energy of the n photons $n\omega$ is absorbed and converted into the energy $\omega_p = \kappa^2/2$ needed to detach the electron from the atom, the kinetic energy $\vec{p}^2/2$ of the free electron as well as the ponderomotive energy. The function $F_n(\vec{p})$ corresponds to a momentum-resolved photo-electron spectrum of the n-photon process, and its modulus can experimentally be measured. With the substitutions

$$r = (1 + (p_y^2 + p_z^2)/\kappa^2), \quad \gamma' = \gamma r, \quad \omega' = \omega/r^2, \quad p' = p_x/r \qquad (4.23)$$

it follows that

$$F_n(\vec{p}) = \frac{i^n}{2\pi} \int\limits_{-\pi}^{\pi} d\beta \chi_{\ell m}(\pi(\beta)) \exp\left\{-i\frac{\omega_p}{\omega'}\left[\left(\frac{p'^2}{\kappa^2}+1+\frac{1}{2\gamma'^2}\right)t+\frac{2p'}{\kappa\gamma'}\sin\beta+\frac{1}{4\gamma'^2}\sin 2\beta\right]\right\}$$

$$(4.24)$$

where $\chi_{\ell m}(\vec{\pi}(\alpha) = 1/2(\vec{\pi}^2(\alpha) + \kappa^2)\tilde{\phi}_{\ell m}(\vec{\pi}(\alpha))$, and $\tilde{\phi}_{\ell m}(\vec{p})$ is the momentum representation of the bound state electronic wavefunction which is obtained from the configuration space wavefunction by a 3D Fourier transform. In order to obtain an approximate numerical exprefigure ssion for the integral, some further approximations are made by the authors of [10]. First, the saddle point approximation is used to evaluate the integral, where it is presupposed that the exponential provides a rapidly oscillating function. Then, only stationary points of the exponent contribute to the integral. Second, it is shown that, in the space domain, the wavefunction $\phi_{\ell m}$ of the bound electronic state needs to be known asymptotically only, i.e., only values $\phi_{\ell m}(\vec{r})$ for $r \to \infty$ enter the integral Eq. (4.24) evaluated according to the saddle point approximation. Third, it has to be assumed that the electron wavefunction of the initial state corresponds to that of a bound state in a potential that vanishes more rapidly than the Coulomb potential $\sim 1/r$. Therefore, the rate finally obtained in Ref. [10] is strictly valid only for, e.g., the photo-detachment of electrons from negatively charged ions, and is given by

$$w_{\ell m}(\gamma, \omega) = \frac{4\sqrt{2}}{\pi}\omega_p C_{n_*\ell_*}^2 \frac{(2\ell + 1)(\ell + |m|)!}{2^{|m|}|m|!(\ell - |m|)!}\frac{\gamma^2}{1 + \gamma^2}A_m(\gamma)$$

$$\times \left(\frac{\Theta}{E\sqrt{1 + \gamma^2}}\right)^{-3/2} \exp\left[-\frac{\Theta}{3E}g(\gamma)\right], \qquad (4.25)$$

where all quantities have been converted back to SI units. The internal atomic field strength is given by $\Theta = \frac{4}{q}\omega_p^{3/2}\sqrt{2m_e\hbar}$. The coefficient $C_{\ell_*n_*}^2$ results from an asymptotic expansion of the electronic wavefunction of the atoms and is given by

$$C_{\ell_*n_*}^2 = \frac{2^{2n_*}}{n_*\Gamma(n_* + \ell_* + 1)\Gamma(n_* - \ell_*)} \qquad (4.26)$$

In fact, for atomic hydrogen, the asymptotic expansion of the wavefunction is known exactly, and one may set $\ell_* = \ell$ and $n_* = n$, where n and ℓ are the principal and orbital quantum numbers, respectively. For more complex atoms, an approximate expression for these coefficients can be obtained by introducing effective quantum numbers according to $n_* = Z\sqrt{\omega_H/\omega_P}$ [29] and $\ell_* = n_* - 1$ [30, 31], where ω_p and ω_H correspond to the ionization potential of the respective atom and atomic hydrogen, respectively. The function $A(\gamma)$ can be expressed as the following infinite series:

$$A_m(\gamma) = \frac{1}{|m|!} \sum_{\kappa \geq \nu}^{\infty} e^{-\alpha(\kappa-\nu)} w_m\left(\sqrt{\beta(\kappa-\nu)}\right) \qquad (4.27)$$

Here, the following notation is used:

$$w_m(x) = e^{-x^2} \int_0^x e^{t^2} (x^2 - t^2)^{|m|} dt \qquad (4.28)$$

$$\alpha = 2\left[\sinh^{-1}\gamma - \frac{\gamma}{\sqrt{1+\gamma^2}} \right] \qquad (4.29)$$

$$\beta = \frac{2\gamma}{\sqrt{1+\gamma^2}} \qquad (4.30)$$

$$\nu = \frac{\omega_p}{\omega}\left(1 + \frac{1}{2\gamma^2}\right) \qquad (4.31)$$

$$\kappa = \langle \nu + 1\rangle + k, \quad k = 0, 1, 2, 3, \dots \qquad (4.32)$$

In the definition of κ, $\langle x\rangle$ denotes the integer part of x. Under experimental conditions typically met in femtosecond filamentation in noble gases, it is reasonable to assume that the orientation of all atoms is equally distributed. Thus, in the following, it suffices to consider the ionization rate $w_\ell = \frac{1}{2\ell+1}\sum_{-\ell}^{\ell} w_{\ell m}$, averaged over all possible magnetic quantum numbers m. One then obtains the simplified expression

$$w(\gamma, \omega) = \frac{4\sqrt{2}}{\pi}\omega_p C_{\ell_* n_*}^2 \frac{\gamma^2}{1+\gamma^2} A_0(\gamma)\left(\frac{\Theta}{E\sqrt{1+\gamma^2}}\right)^{-3/2} \exp\left[-\frac{\Theta}{3E}g(\gamma)\right] \qquad (4.33)$$

with $\ell_* = n_* - 1$, and it has been used that $A_m \ll A_0$ for $m \neq 0$.

Coulomb Corrected Ionization Rates

The ionization rate Eq. (4.33) was derived under the assumption of a short range atomic potential. This neglects the Coulomb interaction of the ionized electron with the atomic residuum and does not adequately describe the ionization of neutral atoms. In Ref. [28], dressed continuum states, i.e., Coulomb corrected Volkov functions, were used to account for long-range interactions. An alternative approach is provided by the method of imaginary times [32]. This method treats the sub-barrier motion of the tunneling electron quasiclassical, i.e., it calculates classical electron trajectories according to the Newton's equations of motion. However, conservation of energy in Newtonian mechanics involves the relation $dt = dr/\sqrt{2(E - V)/m}$, where E is the conserved energy and V the barrier potential. It is thus clear that sub-barrier movement ($E < V$) can classically be treated only in imaginary time. Having solved for the classical trajectories of electron tunneling with appropriate initial- and boundary conditions, the Coulomb correction to the ionization rate is essentially determined by the classical action integral $S = \int dt\, L$, where $L = T - V$ is the Lagrangian of an electron under the influence of both the Coulomb potential and the external alternating electric field, and T denotes the kinetic energy. In fact, this method is closely related to similar methods used in semiclassical theories, as for example the well-known WKB (Wenzel, Kramers, Brillouin) approximation of quantum mechanics. Furthermore, it was shown in Ref. [33] that, using the saddle-point approximation, the path-integral representation of the quantum-mechanical Greens function may be approximated as

$$G(\vec{r}, t; \vec{r}', t') \approx \frac{\theta(t - t')}{(2\pi i(t - t'))^{3/2}} \exp\left(iS(\vec{r}, t; \vec{r}', t')\right) \tag{4.34}$$

where $S(\vec{r}, t; \vec{r}', t')$ is the action evaluated along a classical trajectory connecting the points $(\vec{r}, t)$ and $(\vec{r}', t')$. Also, corresponding semiclassical method may be used in attosecond physics to describe the recollision dynamics of electrons relevant in the generation of attosecond pulses and high harmonic radiation [34, 35]. The method of imaginary times was used 1967 by Popov et al. [36] to derive a Coulomb corrected version $w_C(\gamma, \omega)$ of their original ionization rate $w(\gamma, \omega)$ Eq. (4.25). They obtained

$$w_C(\gamma, \omega) = Q w(\gamma, \omega), \quad Q = \left(\frac{\Theta}{E}\right)^{2n_*}. \tag{4.35}$$

The correction factor Q strongly increases the ionization rates, as the potential barrier generated by the superposition of a long-range Coulomb potential is strongly suppressed compared to the case involving a short-range atomic potential.

4.2.2 A Recent Modification of the PPT Model

Recently, an improved ionization rate was derived in Ref. [7] according to

$$w(\gamma, \omega) = \omega_p \frac{2^{2n^*-2}}{\Gamma^2(n^*+1)} \left(\frac{\omega_p}{\omega}\right)^{-3/2} \beta^{1/2} A_0(\gamma) \left(\frac{\Theta}{E(1 + 2e^{-1}\gamma)}\right)^{2n^*} \exp\left[-\frac{\Theta}{3E}g(\gamma)\right] \tag{4.36}$$

where the definition of the quantities $A_0(\gamma), g(\gamma), \beta$ matches that of Eq. (4.25). Again, the method of imaginary time was used to derive a Coulomb corrected ionization rate. However, unlike the derivation of Ref. [36], the authors laid special emphasis on the perturbative limit $\gamma \gg 1$ and succeeded in presenting a very accurate Coulomb corrected ionization rate in this regime. Thus, this result is especially useful for intense, high-frequency XUV or x-ray laser radiation with $\gamma \gg 1$, as produced, e.g., by free electron lasers. Nevertheless, the latter results were shown to be equally valid in the tunneling regime $\gamma \ll 1$ and, in fact, also for arbitrary frequencies and values of the Keldysh parameter γ. As the subsequent analysis strongly relies on the accuracy of calculated MPI cross-sections, in the following, Eq. (4.36) is used as a starting point for derivation of the latter.

4.2.3 The Multiphoton Limit

In the following, the limit of expression Eq. (4.36) in the perturbative regime $\gamma \gg 1$ is calculated, for which

$$
\begin{aligned}
g(\gamma) &\to \tfrac{3}{2}\gamma^{-1}(\ln 2\gamma - \tfrac{1}{2}), \\
\alpha(\gamma) &\to 2(\ln 2\gamma - 1), \\
\beta(\gamma) &\underset{\gamma \to \infty}{\to} 2.
\end{aligned}
\tag{4.37}
$$

In addition, for large γ one has

$$
\left(\frac{\Theta}{E(1 + 2e^{-1}\gamma)}\right)^{2n^*} \to \left(\frac{2e\omega_p}{\omega}\right)^{2n^*},
\tag{4.38}
$$

$$
\exp\left[-\frac{\Theta}{3E}g(\gamma)\right] \to \exp\left[-\frac{\Theta}{3E}\frac{3}{2}\gamma^{-1}\left(\ln 2\gamma - \frac{1}{2}\right)\right] = (2\gamma)^{-2\omega_p/\omega}\exp\left(\frac{\omega_p}{\omega}\right)
\tag{4.39}
$$

It follows that for $\gamma \gg 1$, the infinite sum represented by $A_0(\gamma)$ reduces to

$$
A_0(\gamma,\omega) \to \sum_{k=0}^{\infty}(2\gamma)^{-2(K+k-\nu_p)}e^{2(K+k-\nu_p)}w_0\left(\sqrt{2(K+k-\nu_p)}\right),
\tag{4.40}
$$

where $K = \langle\omega_p/\omega + 1\rangle$ is the minimum number of photons required to ionize an atom with ionization potential $\hbar\omega_p$ and $\nu_p = \omega_p/\omega$. With Eqs. (4.38)–(4.40) and the definition of γ in Eq. (4.10), the ionization rate (4.36) admits the following perturbative expansion in the limit $\gamma \gg 1$,

$$
w = \sum_{k=0}^{\infty}\sigma_{K+k}I^{K+k},
\tag{4.41}
$$

where σ_k, the cross-section for k-photon absorption, is defined below. Furthermore, with the definition Eq. (2.32), the Keldysh parameter Eq. (4.16) was expressed in terms of the intensity I. The linear refractive index n_0 appearing in the definition of the intensity was set equal to unity, $n_0 = 1$. This is justified by that fact that only gases at standard conditions are considered in the following.

The leading term of the expansion corresponds to K-photon ionization with the minimum number of photons required, whereas the higher order terms describe above threshold ionization (ATI) [37], which results in an excess kinetic energy of the ionized electron that is larger than the ionization energy ω_p as more photons are absorbed than required for ionization. However, as the PPT model fully neglects any internal atomic resonances, effects such as resonance-enhanced MPI [38] where the atom is excited to an intermediate bound state from which it is subsequently ionized, or so called rescattering processes [37], where the electron returns to the ionic residuum and is scattered off it, are not accounted for in this model. The MPI cross-sections σ_K derived from the perturbative limit of the ionization rate of [7] is given by

$$\sigma_K = \frac{2\sqrt{2}C^2}{\pi} 2^{2n^*-2K} e^{2n^*} \omega_p \left(\frac{\omega_p}{\omega}\right)^{2n^*-3/2}$$

$$\times e^{2K-\omega_p/\omega} \omega^{-2K} \left(\frac{q^2}{\hbar\omega_p m_e \epsilon_0 c}\right)^K w_0\left[\sqrt{2K - 2\frac{\omega_p}{\omega}}\right] \qquad (4.42)$$

Each σ_K is defined for $\omega > \omega_p/K$, where $\omega_p = U_i/\hbar$ is the minimum frequency required for a single photon to ionize the atom. A log–log plot of the spectral dependence of the MPI cross sections σ_2, σ_3 and σ_4 for argon ($U_i = 15.76\,\text{eV}$) according to Eq. (4.42) is shown in Fig. 4.3. The cross-sections exhibit a strongly asymmetric spectral dependence and drop off sharply towards zero at the K-photon absorption edge at ω_p/K.

Having found a perturbative expansion of the ionization rate now lays the ground for an application of KK theory to calculate the coefficient n_{2k} in the perturbative expansion of the Kerr refractive index $\Delta n = n_2 I + n_4 I^2 + \cdots$ from the coefficients σ_k. This will be the aim of the following sections.

4.3 Kramers-Kronig Approach to Second Order Nonlinear Refraction

Second order nonlinear refraction is a key nonlinear optical mechanism in isotropic media, including all gases, liquids, and a large class of solids. In dielectric media, nonlinear refraction causes an intensity-dependent increase of the index of refraction $n = n_0 + n_2 I$, which gives rise to spectral broadening and is the basis for nearly all femtosecond pulse compression mechanisms. While nonlinear refraction in solids and liquids has been directly linked to two-photon absorption via a modified KK relationship [8, 9, 39] and has been extensively explored experimentally using the z-scan technique [40], neither one of these can be used to determine the nonlinear

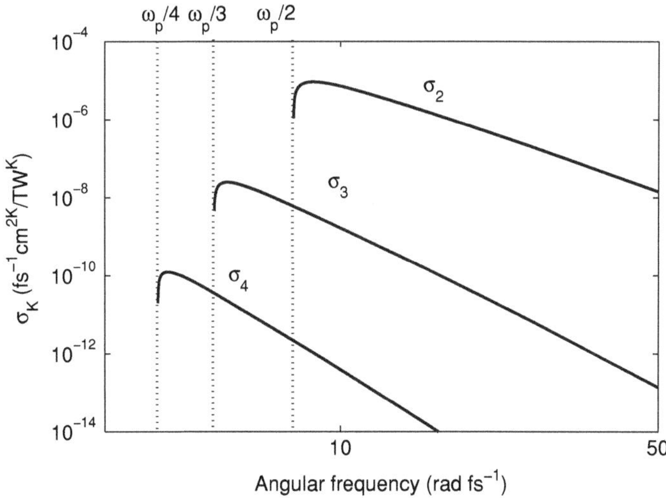

Fig. 4.3 Log–log plot of cross section σ_2 to σ_4 of multiphoton ionization as derived from Eq. (4.42)

index of refraction n_2 in gases. Despite the high technical importance of this nonlinear spectral broadening mechanism in gaseous media [41–43], all theoretical modeling of the latter therefore relies on an indirect determination of $n_2 \propto \gamma^{(3)}(-\omega; \omega, \omega, -\omega)$ from third-harmonic generation (THG) measurements [44] $\gamma^{(3)}(-3\omega; \omega, \omega, \omega)$ or theoretical calculations of the dynamic hyperpolarizability [45, 46] $\gamma^{(3)}(-2\omega; \omega, \omega, 0)$, which mostly refer to the scenario of electric-field induced second harmonic generation (ESHG). Here, the dynamical hyperpolarizability $\gamma^{(3)}$ is related to the nonlinear optical susceptibility $\chi^{(3)}$ via the Lorentz-Lorenz law[47, 48]. This law emerges from a comparison of the local atomic or molecular dipole moments and the macroscopic polarization P and can easily be generalized to the nonlinear optical case. In this case, the higher-order expansion coefficients relating the **local** electric field to the nonlinear polarization are termed hyperpolarizabilities. Generalizing the Lorentz Lorenz law, the third-order hyperpolarizability $\gamma^{(3)}$ is related to the nonlinear susceptibility $\chi^{(3)}$ according to

$$\chi^{(3)}(-\omega_T; \omega_1, \omega_2, \omega_3) = \frac{\rho_0}{3!\epsilon_0}\left(\frac{\epsilon(\omega_T)+2}{3}\right)\left(\frac{\epsilon(\omega_1)+2}{3}\right)\left(\frac{\epsilon(\omega_2)+2}{3}\right)$$
$$\times \left(\frac{\epsilon(\omega_3)+2}{3}\right)\gamma^{(3)}(-\omega_T; \omega_1, \omega_2, \omega_3) \qquad (4.43)$$

for the case of the third-order hyperpolarizability. Here, $\omega_T = \omega_1 + \omega_2 + \omega_3$, and ϵ denotes the relative permittivity.

Even though the efficiency of second harmonic generation in an isotropic medium in the presence of a strong constant electric field is very accurately measurable,

Table 4.1 Nonlinear refractive index n_2 at 800 nm and atmospheric pressure

$n_2(10^{-8}\mathrm{cm}^2/\mathrm{TW})$	Eq. (4.46)	[44]	Refs. [45, 49]
He	0.52 (0.48)	0.40	0.40 (0.38)
Ne	1.31 (1.18)	0.71	0.99 (0.96)
Ar	12.68 (10.84)	9.46	11.2 (10.4)
Kr	30.69 (25.63)	25.9	25.6 (23.17)
Xe	91.58 (73.87)	77.0	69.8 (61.39)

ESHG data has been compiled from Refs. [45, 49] and corrected for the dispersion of the DFWM process [Eq. (4.51)]. Data from [44] is scaled from $\lambda = 1055$ nm using Eq. (4.49) with $\nu = 3/2$. Values in parentheses denote the static limit $n_2(\omega \to 0)$.

the wavelength dependence of the ESHG coefficients $\gamma^{(3)}(-2\omega; \omega, \omega, 0)$ differs from the dispersion of the coefficients $\gamma^{(3)}(-\omega; \omega, \omega, -\omega)$ governing nonlinear refraction [49]. Equality of the two coefficients is only expected for the limiting case $\omega \to 0$, i.e., in the far infrared. The most accepted experimental data for n_2 have probably been measured by Lehmeier et al., determining the THG efficiency in the inert gases [44]. Because this data has only been determined at the wavelength $\lambda = 1.055\,\mu$m, frequency scaling is generally considered difficult. For argon, using Eq. (18) of Ref. [44] yields $n_2 = 1.33 \times 10^{-19}\,\mathrm{cm}^2/\mathrm{W}$ at 248 nm, which disagrees with independently measured values $n_2 = 2.9 \pm 1.0 \times 10^{-19}\,\mathrm{cm}^2/\mathrm{W}$ of Ref. [50]. Both values, finally, appear to be incompatible to explain the high efficiency of a hollow fiber compressor at 248 nm, which indicates an even higher value of n_2 at this wavelength [51]. This example makes it clear, that there is urgent need for improved scaling laws and more dependable theoretical estimates of the latter. Consequently, and probably owing to the wide spread of experimental data on n_2 published [44, 50–52], values used for the modeling typically vary over an order of magnitude, even for the most commonly used inert gases.

In the following, a different approach is described to deduce n_2 in the inert gases from Keldysh theory [10, 19, 53–55] using the modified KK relationship from Ref. [8]. However, it is important to point out that cross sections for multiphoton ionization are available only for the degenerate case, i.e., the absorption of photons of equal frequency. Therefore, KK theory cannot be used to relate degenerate absorption coefficients for TPA to self-refraction $n_2(\omega) = \frac{3}{4n_0^2\epsilon_0 c}\chi^3(-\omega; \omega, \omega, -\omega)$. Instead, one introduces $n_2^{(ND)}(\omega_1, \omega_2) = \frac{3}{2n_0^2\epsilon_0 c}\chi^3(-\omega_1; \omega_1, \omega_2, -\omega_2)$ which describes the refractive index change seen by a probe beam at ω_1 induced by the presence of a strong pump beam at ω_2. Here, a factor of 2 arises due to the so-called weak-wave retardation [56]. This states that generally, nonlinear effects induced by auxiliary beams are a factor of 2 higher than the corresponding self-action effects. Then, for the nondegenerate $n_2^{(ND)}$, a KK relation between $n_2(\omega_1, \omega_2)$ and the cross section for nondegenerate TPA $\Delta\alpha^{(ND)}(\omega_1, \omega_2)$ can be formulated as

$$\Delta n(\omega_1, \omega_2) = n_2(\omega_1, \omega_2)I = \frac{c}{\pi}\mathcal{P}\int_0^\infty \frac{\Delta\alpha^{(ND)}(\Omega, \omega_2)}{\Omega^2 - \omega_1^2}d\Omega \qquad (4.44)$$

where $\mathcal{P}$ denotes Cauchy's principal value. Unfortunately, there exist neither theoretical nor experimental work on the dispersion of the cross sections for nondegenerate TPA or multiphoton absorption in rare gases. Thus, one has to use an estimate which also has been used successfully in Ref. [8] for the calculation of nonlinear refraction in solids from TPA coefficients using KK theory. With $\Delta\alpha(\omega)$ being the TPA coefficient for the degenerate case, in the latter work, the estimate $\Delta\alpha^{(ND)}(\omega_1, \omega_2) = 2\Delta\alpha^D((\omega_1 + \omega_2)/2)$, was used, with a factor of 2 arising due to the weak-wave retardation. Of course, the latter approximation is only reasonable for $\omega_1 \approx \omega_2$. However, the approximation is justified by noting that the presence of the denominator in Eq. (4.44) strongly weights frequencies Ω in the vicinity of ω_1. Indeed, a remarkable agreement with measured values of the nonlinear refraction in solids was obtained in Ref. [8]. The TPA absorption coefficient $\Delta\alpha_2(\omega)$ can be related to the TPA cross section $\sigma_2(\omega)$ (for the degenerate case) according to [57]

$$\Delta\alpha(\omega) = 2\hbar\omega\rho_0\sigma_2(\omega)I. \qquad (4.45)$$

Thus, with the above considerations, setting $\omega_1 = \omega_2 = \omega$ in Eq. (4.44), a KK relation for self-refraction may be formulated as

$$n_2(\omega) = \frac{\hbar c\rho_0}{\pi}\mathcal{P}\int_0^\infty \frac{\sigma_2\left(\frac{1}{2}(\Omega + \omega)\right)}{\Omega - \omega}d\Omega. \qquad (4.46)$$

In order to evaluate integrals of the form of Eq. (4.46), here a method based on the Fast Fourier Transform (FFT) is used. This is possible as it turns out that the integral transform in Eq. (4.46) is closely related to the Hilbert transform (HT) $\mathcal{H}$. The HT of a function f is defined as

$$\mathcal{H}[f](\omega) = -\frac{1}{\pi}\mathcal{P}\int_{-\infty}^\infty \frac{f(\Omega)}{\Omega - \omega}d\Omega \qquad (4.47)$$

This can be evaluated by employing the following formula which relates the HT $\mathcal{H}$ to the Fourier transform $\mathcal{F}$ according to [58]

$$\mathcal{F}\big[\mathcal{H}[f]\big](\omega) = -i\,\mathrm{sgn}(\omega)\mathcal{F}[f](\omega) \qquad (4.48)$$

Numerically, this is evaluated using the FFT method. Equation (4.46) then yields the dispersion of n_2 with wavelength for the five inert gases helium, neon, argon, krypton and xenon. The results of these calculations in comparison to independent experimental and theoretical data are shown in Fig. 4.4. Experimental data has been mainly extracted from measurements of the third-harmonic efficiency by Lehmeier et al. [44], which is probably the most accepted reference on experimental data for n_2 of the inert gases. As all data has been acquired at a wavelength $2\pi c/\omega' = 1055\,\text{nm}$, it is extrapolated to a range of angular frequencies ω shown in Fig. 4.4 to indicate the dispersion of n_2 with wavelength. For this purpose, the relation

$$n_2(\omega) = \frac{\nu\omega' - \omega_p}{\nu\omega - \omega_p} n_2(\omega'),\tag{4.49}$$

originally suggested in Ref. [44] to scale THG data with $\nu = 3$ is used, keeping in mind that in the theoretical computations by Bishop and Pipin [49], the DFWM coefficient $\gamma^{(3)}(-\omega; \omega, \omega, \omega)$ scales differently with wavelength than its THG counterpart $\gamma^{(3)}(3\omega; \omega, \omega, \omega)$. The data was augmented by theoretical calculations [46, 49] and measurements [45] of the hyperpolarizability of the gases under consideration. With the noted exception of the data provided in Ref. [49], however, the hyperpolarizability data only provide ESHG efficiencies, which unfortunately exhibit a different dispersion behavior. Without further processing, this data can therefore only be sensibly used to estimate the long wavelength limit $n_2(\omega \to 0)$. Based on the extensive work of Ref. [49], this issue can be fixed by noting that in the long-wavelength limit, the third-order hyperpolarizability may be expanded into an even power series according to

$$\gamma^{(3)} = \gamma_0(1 + A\omega_L^2 + B\omega_L^4 + \cdots)\tag{4.50}$$

with $\gamma_0 = \gamma^{(3)}(0; 0, 0, 0)$. The coefficient A has been demonstrated in Ref. [59] to be the same for all third-order nonlinear processes, and the difference in frequency scaling is governed by the quantity $\omega_L^2 = \nu\omega^2$, where $\nu = 12, 6$ and 4 for THG, ESHG and DFWM, respectively.

As $\nu_{\text{DFWM}} : \nu_{\text{ESHG}} = 2 : 3$, it is obvious to rescale ESHG data according to

$$\gamma_{\text{DFWM}}(\omega) \approx \gamma_0 \left[\frac{\gamma_{\text{ESHG}}(\omega)}{\gamma_0} \right]^{2/3}\tag{4.51}$$

in order to obtain the hyperpolarizability γ_{DFWM} associated to degenerate four-wave mixing. Again, this adjustment provides a much better agreement with the experimental data for gases like krypton and argon at 248 nm wavelength [50] but fails to explain the values reported for neon and the large negative n_2 of xenon, which is, however, attributed to a local coincidence with a two-photon resonance. The main reason for including the ESHG data is their much higher reliability. Typically, a precision of about 2 % or better is claimed for ESHG data, whereas values of 10 % are already considered as extremely reliable for all methods of determining hyperpolarizabilities with three optical fields, i.e., DFWM or THG measurements. Inspecting the data in Fig. 4.4, one finds that in the long wavelength limit all different methods for estimating n_2 agree with each other within 20–30 %. Considering both, the approximations made by PPT theory, and the assumptions used here on the

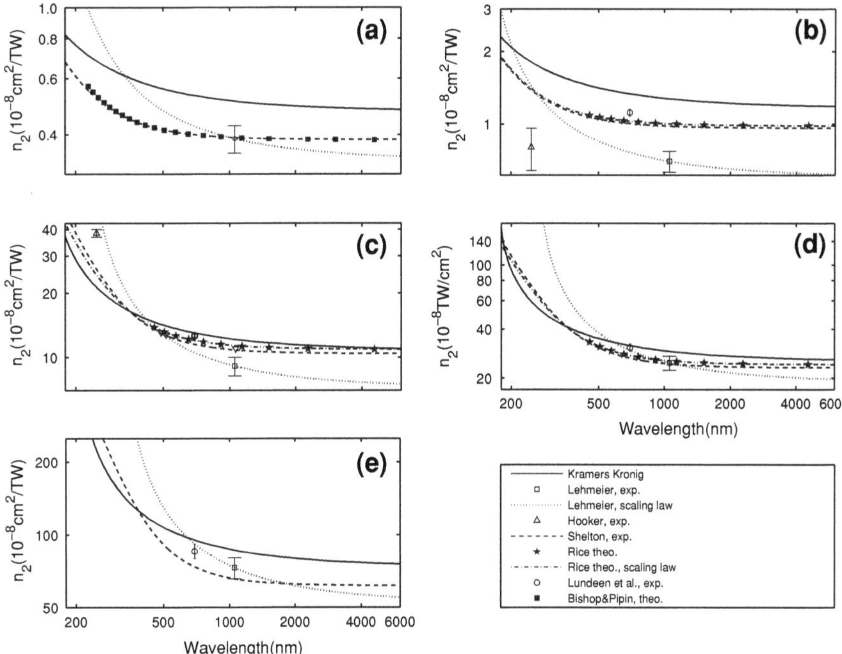

Fig. 4.4 Nonlinear refractive index in the long-wavelength regime according to Eq.(4.46) for helium (**a**), neon (**b**), argon (**c**), krypton (**d**) and xenon (**e**). *Squares* experimental data of Lehmeier et al. [44]. *Triangles* Hooker et al., experiment [50]. *Stars* Rice et al., theory [46]. *Circles* experimental data, Lundeen et al. [60]. *Filled squares* Bishop and Pipin, theory [49]. *Solid lines* n_2 according to Eq.(4.46). *Dashed lines* experimental data of Ref.[45], rescaled using Eq.(4.51). *Dashed-dotted lines* theoretical calculations of Ref.[46], interpolated according to scaling law Eq.(4.50). *Dotted lines* Lehmeier data [44] rescaled with the help of Eq.(4.49)

frequency dependence of the nondegenerate TPA absorption coefficient, Eq.(4.46) delivers excellent n_2 estimates in the infrared and visible, with a slight tendency to underestimate nonlinear refraction compared to measured hyperpolarizabilities and ESHG data. Disagreement between ESHG data and the KK expansion in the infrared limit $\omega \to 0$ is typically only 10–20 %. Compared to Ref.[11], which neglected the frequency dependence of the TPA cross-section, the KK expansion yields good agreement with independent theoretical and experimental work also below 500 nm. Table 4.1 lists a compilation of n_2 valuesLWL for 800 nm, which is currently probably the most important wavelength for pulse compression experiments. Keeping in mind that the absorption spectra σ_K have been derived from strong field ionization rates for which often an order of magnitude agreement with experimental data is considered reasonable, the KK approach provides an outstanding agreement with the n_2 values provided by the cited theoretical and experimental work. Especially, considering the static limit $n_2(\omega \to 0)$, the values calculated from the KK approach deviate no more than 15 % from the cited experimental and theoretical reference.

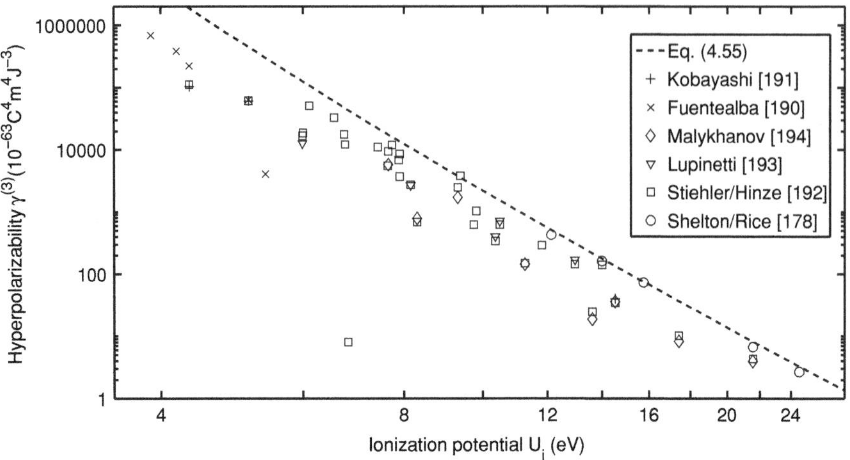

Fig. 4.5 Static limit $\gamma^{(3)}(0; 0, 0, 0)$ for various atoms versus ionization potential. The data provided by Ref. [45] was obtained experimentally, while Refs. [61–65] provide theoretically calculated for the hyperpolarizability. For comparison, the *dashed line* depicts the values obtained from KK theory using Eq. (4.53)

For the static limit $\gamma^{(3)}(0; 0, 0, 0)$ of the hyperpolarizability, a completely analytical expression may be obtained by relating the hyperpolarizability to the Kerr coefficient n_2 with the help of Eqs. (4.43) and (2.50) which yields

$$\gamma^{(3)}(0) = \frac{8\epsilon_0^2 c}{\rho_0} n_2(0), \tag{4.52}$$

where it was used that $n_0 = \sqrt{\epsilon} \approx 1$. Evaluating Eq. (4.46) in the limit $\omega \to 0$ finally yields

$$\gamma^{(3)}(0) = \frac{64 e^4 \hbar^2 \epsilon_0^2}{\omega_H^4 m_e^3} F(n_*) \tag{4.53}$$

where the function $F(n_*)$ describes scaling of n_2 with the effective principal quantum number $n_* = \sqrt{\omega_H/\omega_p}$ of the gas species and is given by the integral representation

$$F(n_*) = \frac{(8e)^{2n_*}}{\Gamma^2(n_* + 1)} n_*^{10} \int_0^1 dx\, x^{2n_*+3/2} e^{-2x} w_0\left[2\sqrt{1-x}\right] \tag{4.54}$$

The variation of $\gamma^{(3)}(0)$ with the atomic ionization potential is shown as dashed line in Fig. 4.5.

For comparison, the figure is supplemented with the static hyperpolarizabilities of various atomic species. All data was theoretically obtained by different perturbative techniques [61–65] except for hyperpolarizabilities of the rare gases He, Ne, Ar, Kr, Xe which were taken from experimental work [45].

Clearly, for the noble gases, the agreement with the theoretical curve is most striking. However, for decreasing ionization potential, the n_2 values found in the literature follow only roughly the curve calculated here, with an increasing tendency of the present theory to overestimate the static n_2 values. However, it should be kept in mind that the PPT model only provides a rough estimate on the MPI cross sections σ_K, as this model neglects any internal atomic resonances. Rather, only transitions from the bound state to the continuum are considered. Thus, processes such as resonance enhanced MPI [37, 38] or multiphoton transitions between bound states [66], are not accounted for by the cross section σ_K derived from PPT theory. For example, considering two-photon bound-bound resonances in xenon, the authors of [66] predict negative n_2 for xenon at 248 nm. Nevertheless, the excellent agreement of the static n_2 values of the noble gases mainly considered here, suggests that the dominant contribution to the second-order nonlinear refractive index results from non-resonant TPA due to transitions between a bound state and the continuum.

In conclusion, an alternative route for estimating the nonlinear refraction of inert gases from a KK transformation of the PPT theory has been established. This method is completely analytical and requires knowledge of a single parameter, i.e., the ionization energy of the gas. Equation (4.46) directly delivers estimates that are in surprisingly good agreement with experimentally measured values for the visible and infrared. As it will be shown in the next section, this method can easily be generalized for the estimation of higher-order effects or other gases where less experimental information is available. Moreover, the dispersion of n_2 at frequencies $\omega > \omega_p/2$ above the two-photon absorption edge has been completely disregarded. This also will be supplemented in the next section.

4.4 Higher Order Kerr Effect and Femtosecond Filamentation

In this section, the KK ansatz will be generalized to calculate higher-order indices. This finally provides an independent check of the formerly reported saturation behavior of the Kerr refractive index in gaseous media. Keeping in mind the power series expansion Eq. (4.1) of the nonlinear refractive index, it is obvious that $n_{2(K-1)}$ is related to K-photon absorption, as the absorption coefficient scales with intensity according to [57]

$$\Delta\alpha_K(\omega) = \beta_K I^{K-1}. \tag{4.55}$$

where $\beta_K = K\hbar\omega\rho_0\sigma_K$, cf. Eq. (2.51) which relates β_K to the imaginary part of the nonlinear susceptibility $\chi^{(2K-1)}$. In principle, the KK approach is only applicable to the nondegenerate case of simultaneous absorption of K photons of different frequencies $\omega_1, \omega_2, ..., \omega_K$. In the present case, only degenerate absorption coefficients are known for which no KK relations may be formulated. Thus, analogous to the considerations that led to Eq. (4.44), an estimate has to be provided for the

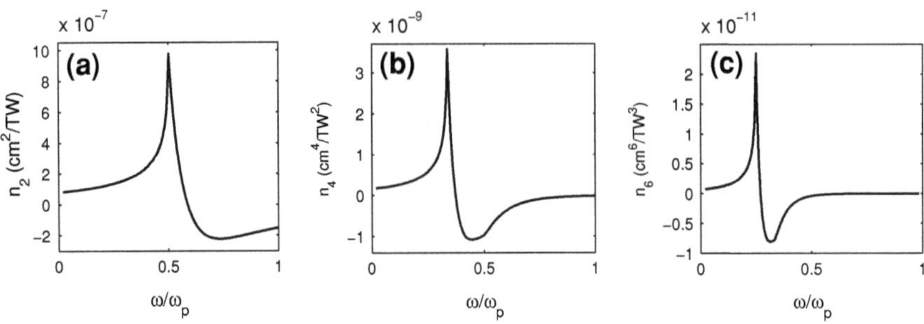

Fig. 4.6 Dispersion curves for $n_2...n_6$ for argon in the vicinity of the 2,3 and 4-photon resonance

nondegenerate absorption cross-section $\sigma_K^{(ND)}(\omega_1, \omega_2, ..., \omega_K)$ for the simultaneous absorption of K photons of frequencies $\omega_1, \omega_2, ...\omega_K$. Again, it seems obvious to use

$$\sigma_K^{(ND)}(\omega_1, \omega_2, ..., \omega_K) = \sigma_K\left(\frac{\omega_1 + \omega_2 + ... + \omega_K}{K}\right), \qquad (4.56)$$

where σ_K is given by Eq. (4.42). Then, the change of refractive index seen by a probe beam at ω_1 due the presence of pump beams at $\omega_2, \omega_3, ...\omega_K$ is related via KK to the nondegenerate absorption coefficient according to ($k = K - 1$ is introduced for notational convenience)

$$n_{2k}(\omega_1, \omega_2, ..., \omega_{k+1})$$

$$= \frac{\hbar c \rho_0}{\pi} \mathcal{P} \int_0^\infty (\Omega + \omega_2 + \cdots + \omega_{k+1}) \frac{\sigma_{k+1}\left(\frac{\Omega + \omega_2 + \cdots + \omega_{k+1}}{k+1}\right)}{\Omega^2 - \omega_1^2} d\Omega. \qquad (4.57)$$

In the present work, only self-refraction is of interest, therefore setting $\omega_2 = \omega_3 = ... = \omega_{k+1} = \omega$ yields

$$n_{2k}(\omega) = \frac{\hbar c \rho_0}{\pi} \mathcal{P} \int_0^\infty (\Omega + k\omega) \frac{\sigma_{k+1}\left(\frac{\Omega + k\omega}{k+1}\right)}{\Omega^2 - \omega^2} d\Omega. \qquad (4.58)$$

This equation is evaluated using the HT method already used for the calculation of the dispersion behavior of n_2 in the previous section. The results for n_2 to n_6 for argon are displayed in Fig. 4.6. Focusing the attention to the dispersion of the refractive indices above the threshold for K-photon absorption ($\omega > \omega_p$), it is observed that the nonlinear refractive index is highly dispersive in the vicinity of the K photon absorption edge. Attaining its maximum at ω_p/K, it rapidly decreases for $\omega > \omega_p/K$ and eventually changes sign. In fact, the emergence of negative numerical values of the higher-order refractive indices is a necessary condition to observe saturation, as also discussed in Refs. [1, 2]. However, for argon, a negative

$n_4 = (-0.36 \pm 1.03) \times 10^{-9} \text{cm}^4/\text{TW}^2$ is predicted at 800 nm, whereas Fig. 4.6 suggests that within the framework of the current model, $n_{2(K-1)}$ can attain negative values only for frequencies $\omega > \omega_p/K$ above the K-photon absorption edge. In fact, the KK approach yields $n_4 = 0.239 \times 10^{-9} \text{cm}^4/\text{TW}^2$. Regarding the errorbars provided Ref. [2], the present value of n_4 is compatible with the results of Loriot et al. Within the KK model, negative values of $n_{2(K-1)}$ for argon are expected for $K \geq 11$, as 800 nm is close to an 11-photon resonance at $\lambda = 11 \times 2\pi c/\omega_p \approx 865.4$ nm.

A possible mechanism to explain the saturation behavior of the nonlinear refractive index has been proposed in Ref. [3], taking into account ionization induced depletion of the ground state, giving rise to a reduction of n_2 for the cations. The degree of ionization enters the definition of the multiphoton cross-section Eq. (4.42) via the ionization potential and the effective principal quantum number $n_* = Z\sqrt{\omega_H/\omega_p}$. For Ar^+ cations, the second ionization potential is found to be $U_i = 27.629$ eV. Using this and $Z = 2$ in Eq. (4.46), an estimate for the n_2-value of singly ionized argon atoms can be obtained. At 800 nm, KK theory predicts a value of $n_2 = 6.14 \times 10^{-8} \text{cm}^2/\text{TW}$, a factor of 2 smaller than the corresponding value for neutral argon, cf. Table 4.1. To simulate the saturation behavior due to ground-state depletion, an intensity dependent effective second order nonlinearity $n_{2,\text{eff}}$ can be calculated according to

$$n_{2,\text{eff}}(I) = p n_{2,\text{Ar}^+} + (1-p) n_{2,\text{Ar}} \tag{4.59}$$

where p is the fraction of singly ionized argon atoms. To reproduce experimental conditions of [1, 2], in order to calculate p, the medium is assumed to be ionized by $t_{\text{FWHM}} = 90$ fs pulses with a Gaussian temporal intensity profile given by $I(t) = I_0 \exp(-2t^2/t_p^2)$, where I_0 is the peak intensity and $t_p = t_{\text{FWHM}}/\sqrt{2\ln 2}$. Using the ionization rate Eq. (4.36) $w(I)$ of Ref. [7] for a center wavelength of 800 nm, p is then given by

$$p = 1 - \exp\left(\int_{-\infty}^{t} dt' w(I(t'))\right) \tag{4.60}$$

A plot of $n_{2,\text{eff}}(I)$ versus peak intensity I is shown in Fig. 4.7. The effective second order nonlinearity $n_{2,\text{eff}}$ is nearly constant for intensities up to 150 TW/cm^2, then decreases monotonously until is has reached 50 % of its original value for intensities of the order of 300 TW/cm^2, where the gas is nearly completely ionized. Taking into account that the nonlinear index change is given by $\Delta n = n_{2,\text{eff}}(I)I$, this observed behavior of $n_{2,\text{eff}}$ translates itself into nearly constant slope of $\Delta n(I)$ up to 150 TW/cm^2 and a reduced constant slope for higher intensities. For comparison, the clamping intensity for the given laser- and medium parameters amounts to 81 TW/cm^2. The intensity for which a notable change in the slope of $\Delta n(I)$ occurs is nearly twice as high. It is thus clear that a depletion induced reduction of n_2 plays a minor role in argon filaments. Furthermore, the observed saturation behavior, especially the change of the index sign, cannot be explained by the depletion model based on KK theory.

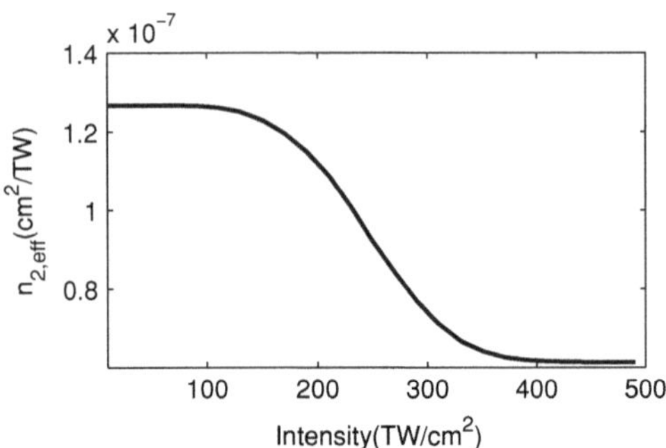

Fig. 4.7 Intensity dependent second order nonlinear refraction $n_{2,\text{eff}}$ in argon at 800 nm according to Eq. (4.59)

Therefore, in order to provide a theoretical model for Kerr saturation and inversion, higher-order nonlinear terms will be included in the definition of the Kerr nonlinear refractive index Eq. (4.1). The higher-order coefficients will be calculated using the KK relation Eq. (4.58) and compared to the experimental results of Refs. [1, 2], which to date provide the only available experimental data on the higher-order nonlinearities. Besides argon, the experimental work of Loriot et al. also provides higher-order Kerr coefficients of the air components N_2 and O_2. However, the model of Ref. [7] only holds to describe photo-ionization of atoms. This is due to the fact that only the asymptotic behavior of the **atomic** initial wavefunction enters the derivation of Eq. (4.36) (cf. Sect. 4.2 of this theses), which therefore fails to properly describe molecule ionization [67]. Instead, more complex approximate approaches using LOPT, cf. Eq. (4.9) and the subsequent discussion, are commonly used to theoretically describe ionization of molecules [18]. In principle, LOPT could be used to calculate higher-order ionization cross-sections σ_K in order to obtain the higher-order Kerr coefficients via the KK transform 4.58. However, the perturbative approach becomes increasingly complex for largefigure K, and closed-form analytical expression for the cross-sections σ_K, comparable to that of Eq. (4.42) for arbitrary K, are not known in the literature. Therefore, in the following the semi-empirical model of Ref. [68] is used, which employs the PPT model with an effective Coulomb potential and an effective residual ion charge Z_{eff}. For O_2 and N_2, the latter work obtains $Z_{\text{eff}} = 0.53$ and 0.9, respectively, which is used in the following to calculate ionization cross-sections σ_K and higher-order Kerr coefficients for the molecular air components. However, for molecular gases, it is known that a delayed Raman response related to the molecular degrees of freedom contributes to the Kerr nonlinearity, cf. Eq. (3.19) and related discussion. It is therefore emphasized that for the molecular gases, Eq. (4.58) only delivers the instantaneous, electronic response to the Kerr nonlinearity, whereas delayed Raman responses are neglected.

The higher-order Kerr coefficients at $\lambda = 800$ nm calculated from the KK transform (4.58) of the ionization cross-sections σ_K are compiled in Table 4.2 for helium, neon, krypton and xenon and for the air components argon, O_2 and N_2. In addition to the n_2-values at 800 nm, for the molecular gases, Eq. (4.53) with the respective Z_{eff} delivers $n_2(0) = 0.7 \times 10^{-7}\,\text{cm}^2/\text{TW}$ and $n_2(0) = 0.8 \times 10^{-7}\,\text{cm}^2/\text{TW}$ for

Table 4.2 Higher-order nonlinearities $n_{2k}(\mathrm{cm}^{2k}/\mathrm{TW}^k)$ for helium, neon, argon, krypton, xenon and molecular oxygen and nitrogen at 800 nm, calculated with Eq. (4.58)

k	He	Ne	Ar	Kr	Xe	O_2	N_2
1	5.21e-09	1.31e-08	1.27e-07	3.07e-07	9.16e-07	8.15e-08	8.80e-08
2	2.41e-12	9.65e-12	2.90e-10	1.09e-09	5.64e-09	3.08e-10	1.92e-10
3	2.48e-15	1.56e-14	1.42e-12	8.27e-12	7.33e-11	2.90e-12	9.19e-13
4	4.54e-18	4.48e-17	1.23e-14	1.11e-13	1.73e-12	5.41e-14	7.94e-15
5	1.31e-20	2.03e-19	1.72e-16	2.45e-15	7.04e-14	1.84e-15	1.11e-16
6	5.54e-23	1.34e-21	3.63e-18	8.62e-17	5.39e-15	1.22e-16	2.39e-18
7	3.23e-25	1.24e-23	1.15e-19	5.14e-18	7.26e-16	9.92e-18	7.82e-20
8	2.52e-27	1.56e-25	5.89e-21	1.22e-18	−4.87e-17	−7.93e-19	4.20e-21
9	2.57e-29	2.59e-27	7.84e-22	−4.28e-20	−6.98e-19	−1.14e-20	8.21e-22
10	3.34e-31	5.78e-29	−3.76e-23	−6.23e-22	−1.05e-20	−1.74e-22	−2.65e-23
11	5.56e-33	1.76e-30	−6.32e-25	−7.75e-24	−1.76e-22	−2.97e-24	−3.61e-25
12	1.19e-34	8.21e-32	−6.77e-27	−1.09e-25	−3.11e-24	−5.33e-26	−3.98e-27
13	3.37e-36	1.20e-32	−8.29e-29	−1.65e-27	−5.64e-26	−9.77e-28	−5.00e-29
14	1.49e-37	−2.43e-34	−1.09e-30	−2.56e-29	−1.03e-27	−1.81e-29	−6.73e-31
15	1.38e-38	−1.92e-36	−1.50e-32	−4.04e-31	−1.91e-29	−3.38e-31	−9.35e-33
16	−3.11e-40	−1.52e-38	−2.08e-34	−6.42e-33	−3.54e-31	−6.31e-33	−1.32e-34
17	−2.09e-42	−1.36e-40	−2.93e-36	−1.03e-34	−6.56e-33	−1.18e-34	−1.88e-36
18	−1.44e-44	−1.30e-42	−4.13e-38	−1.64e-36	−1.22e-34	−2.21e-36	−2.69e-38
19	−1.12e-46	−1.29e-44	−5.86e-40	−2.63e-38	−2.26e-36	−4.13e-38	−3.87e-40
20	−9.37e-49	−1.29e-46	−8.33e-42	−4.23e-40	−4.20e-38	−7.73e-40	−5.56e-42

n_{2k} is related to the cross-section σ_{k+1} of $k + 1$-photon absorption via KK theory.

oxygen and nitrogen, respectively. In fact, the values are in excellent agreement with independent theoretical calculations of Ref. [69] (see also [70]) which provides $n_2(0) = 0.746 \times 10^{-7}\,\mathrm{cm}^2/\mathrm{TW}$ and $n_2(0) = 0.72 \times 10^{-7}\,\mathrm{cm}^2/\mathrm{TW}$ for oxygen and nitrogen, respectively for the purely electronic contribution to the nonlinear refractive index. Furthermore, among the few available references dealing with higher-order susceptibilities of helium, Ref. [71] delivers $n_4 = 2.1 \times 10^{-12}\,\mathrm{cm}^4/\mathrm{TW}^2$, $n_6 = 2.5 \times 10^{-15}\,\mathrm{cm}^6/\mathrm{TW}^3$, $n_8 = 7 \times 10^{-18}\,\mathrm{cm}^8/\mathrm{TW}^4$ and $n_{10} = 4.1 \times 10^{-20}\,\mathrm{cm}^{10}/\mathrm{TW}^5$. These values are in surprisingly good agreement with those resulting from the present KK method.

Interestingly, for all gases considered here, negative $n_{2(K-1)}$ is encountered for $K > U_i/\hbar\omega = \omega_p/\omega$. In this case, the energy of the K absorbed photons is high enough to trigger a multiphoton transition into the continuum. It will now be shown that the appearance of negative Kerr coefficients crucially determines the saturation and inversion behavior of the nonlinear refractive index.

Having computed the nonlinear refractive indices, it is straightforward to calculate the intensity dependent Kerr nonlinear refractive index $\Delta n(I)$ according to Eq. (4.1). For argon, nitrogen and oxygen, the results are presented in Fig. 4.8. For all considered gases, the nonlinear refractive index change Δn exhibits a behavior comparable to theoretical and experimental results of [72, 73, 3, 1, 2], i.e., it saturates and changes sign at intensity levels in the order of magnitude of $\approx 10^{13}\,\mathrm{cm}^2/\mathrm{W}$. The

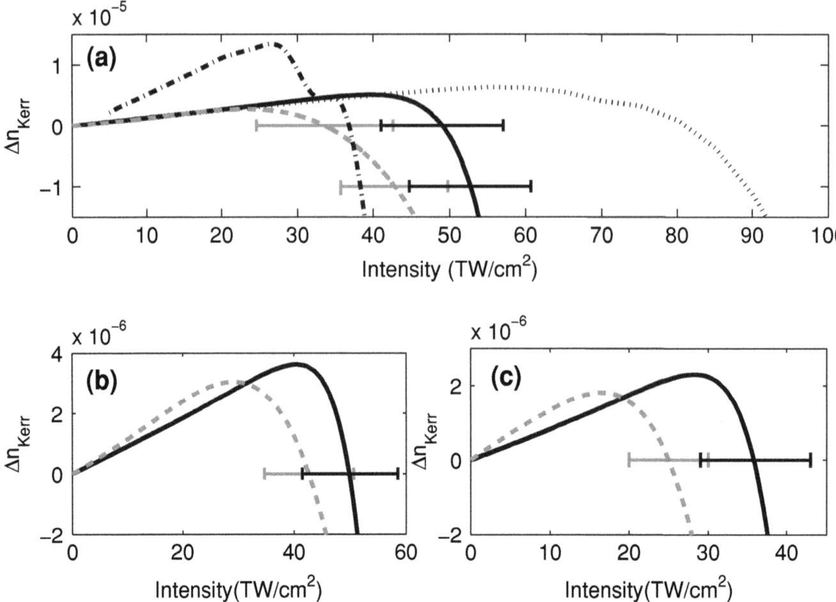

Fig. 4.8 Kerr saturation and inversion in **a** argon, **b** nitrogen and **c** oxygen at 800 nm due to higher order Kerr terms [Eqs. (4.1,4.58), *solid lines*], classical filamentation model due to plasma clamping [Eq. (4.61), *dotted line*], and experimental results [1, 2] (*dashed lines*). *Dashed dotted line* in **a** depicts TDSE results for argon found in [73]. Reprinted from C. Brée et al. [6]. Copyright © 2010 American Physical Society

dashed lines in Fig. 4.8 show experimental results of Ref. [2]. In the latter work, the measured inversion intensity, defined as the (nontrivial) root of $\Delta n(I_{\mathrm{inv}}) = 0$ amounts to 34 TW/cm^2 for argon, whereas the present theoretical results predict a 40 % higher inversion intensity of about 49TW/cm^2. However, Ref. [2] provides error estimates for the measured values $n_2, n_4, ..., n_{10}$. From the numerical values of these errors, it is possible to estimate the error in the experimental inversion intensity of $I_{\mathrm{inv}} = 34 \pm 9\,\mathrm{TW/cm}^2$. Similar considerations hold for the experimental data provided for O_2 and N_2. For the present theoretical results, the error is extracted from the deviation of the lowest order coefficient n_2 from independent data as summarized in Table 4.1, yielding a rough error estimate of ± 20 %. This analysis shows that KK-based calculation of the IDRI yields inversion intensities which favorably agree with experimental results [2]. For argon, an independent prediction of Kerr saturation and inversion was obtained in Ref. [73]. The corresponding $\Delta n(I)$ is shown as the dash-dotted line in Fig. 4.8a. Despite the fact that Ref. [73] slightly overestimates the linear inital slope of $\Delta n(I)$ as determined by the lowest order nonlinear coefficient n_2, the inversion behavior is in good agreement with the present results and that of Loriot et al.

For helium, neon, krypton and xenon, the IDRI $\Delta n(I)$ according to Eqs. (4.58) and (4.1) is plotted versus intensity in Fig. 4.9. Again, for neon, krypton and xenon,

Table 4.3 Inversion intensity from saturation of the nonlinear refractive index versus clamping intensity in the classical model of filamentation

	Helium	Neon	Argon	Krypton	Xenon	O_2	N_2
I_{inv}(TW/cm^2)	113	89	49	40	30	36	50
I_c(TW/cm^2)	301	204	81	57	37	44	82

theoretical results of Ref. [73] are plotted as dash-dotted curves, exhibiting good qualitative agreement with the results provided in this work. The inversion intensities for the gases considered here are summarized Eqs. (4.1)(4.58) in Table 4.3.

In order to elicit the role Kerr inversion assumes in femtosecond filaments, it should be compared to the classical model of filamentation which truncates the Kerr refractive index after the n_2 term and assumes a nonlinear refractive index change

$$\Delta n(I) = n_2 I - \frac{\rho}{2\rho_c}, \tag{4.61}$$

see also the discussion of Sect. 2.5. To reproduce experimental conditions of [1, 2], Eq. 4.61 is evaluated for a 90 fs pulse with a Gaussian temporal profile, for variable peak intensity I. The peak plasma density ρ generated by this pulse is obtained using the ionization model Eq. (4.36) of [7], and n_2 is provided by Eq. 4.46. The results are plotted versus peak intensity I as dotted line in Fig. 4.8a and as dashed lines in Fig. 4.9. The demonstrated behavior of Δn in the clamping intensity model shows clearly that plasma induced saturation and inversion of the nonlinear refractive index occurs at considerably higher intensities as observed within the higher-order Kerr model.

For a quantitative comparison, clamping intensities for the gases under consideration are calculated for a laser wavelength $\lambda = 800$ nm. Instead of using the estimate Eqs. (2.83), (2.81) is solved numerically. The obtained clamping intensities are contrasted with the inversion intensities of the higher-order Kerr model Eq. 4.58 in Table 4.3.

Indeed, the inversion intensities are lower than the clamping intensity. One may thus expect that the saturation of the Kerr refractive index for the considered cases is the dominant mechanism in femtosecond filamentation at 800 nm.

However, it is evident from Eq. 4.58 and Fig. 4.6 that the magnitude of the higher order Kerr coefficients depends on the center wavelength of the irradiated laser beam. It therefore follows that the IDRI and the inversion intensity exhibits dispersion [74], as shown in Fig. 4.10a, b for argon. For comparison, the wavelength dependence of the clamping intensity for argon is shown as dashed line in Fig. 4.10a. The inversion intensity increases towards shorter wavelength, and eventually exceeds the clamping intensity, as independently conjectured in [75]. Therefore, for wavelengths below 600 nm, intensity clamping due to free electrons becomes increasingly relevant and may regain its dominant role for the saturation of the nonlinear refractive index.

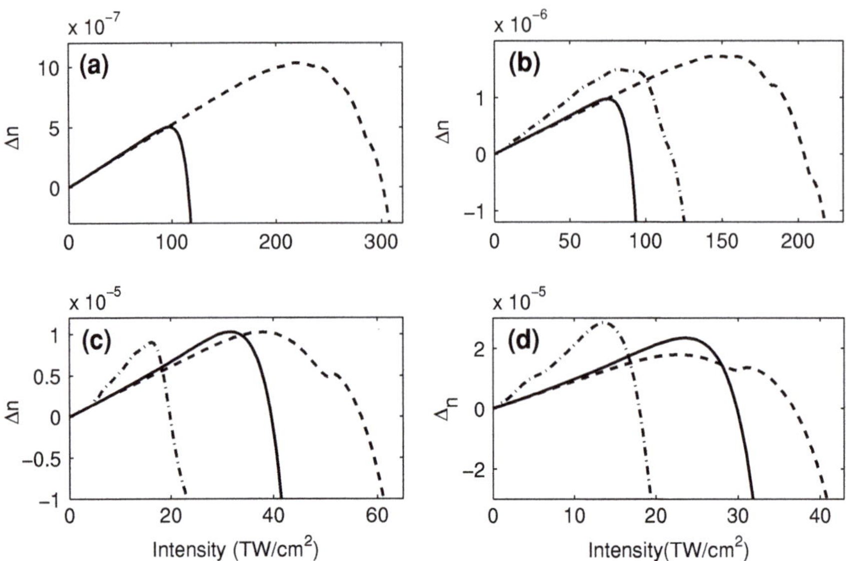

Fig. 4.9   Intensity dependent refractive index for **a** helium, **b** neon, **c** krypton and **d** xenon according to Eq. (4.1) and Table 4.2. *Dashed lines* nonlinear refractive index change in classical model of filamentation, Eq. (4.61). *Dashed-dotted lines* TDSE results of [73]

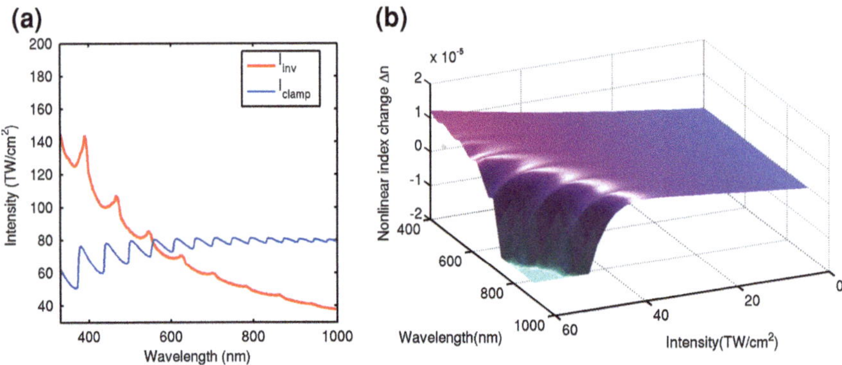

Fig. 4.10   **a** Dispersion of the inversion intensity I_{inv} for argon (*solid line*). *Dashed line* depicts the wavelength-dependent clamping intensity according to Eq. (2.81). **b** Visualization of the dependence of the nonlinear refractive index $\Delta n(I, \omega)$ of argon on wavelength ω and intensity I, calculated according to Eqs. (4.1) and (4.58)

References

1. V. Loriot, E. Hertz, O. Faucher, B. Lavorel, Measurement of high order Kerr refractive index of major air components. Opt. Express **17**, 13429 (2009)
2. V. Loriot, E. Hertz, O. Faucher, B. Lavorel, Measurement of high order Kerr refractive index of major air components: erratum. Opt. Express **18**, 3011 (2010)

3. M. Nurhuda, A. Suda, K. Midorikawa, Generalization of the Kerr effect for high intensity, ultrashort laser pulses. New J. Phys. **10**, 053006 (2008)

4. P. Bejot, J. Kasparian, S. Henin, V. Loriot, T. Vieillard, E. Hertz, O. Faucher, B. Lavorel, J.-P. Wolf, Higher-order Kerr terms allow ionization-free filamentation in gases. Phys. Rev. Lett. **104**, 103903 (2010)

5. M. Kolesik, E.M. Wright, J.V. Moloney, Femtosecond filamentation and higher-order nonlinearities. Opt. Lett. **35**, 2550 (2010)

6. C. Brée, A. Demircan, G. Steinmeyer, Saturation of the all-optical Kerr effect. Phys. Rev. Lett. **106**, 183902 (2011). doi:10.1103/PhysRevLett.106.183902

7. S.V. Poprzuhenko, V.D. Mur, V.S. Popov, D. Bauer, Strong field ionization rate for arbitrary laser frequencies. Phys. Rev. Lett. **101**, 193003 (2008)

8. M. Sheik-Bahae, D.J. Hagan, E.W. van Stryland, Dispersion and band-gap scaling of the electronic Kerr effect in solids associated with two-photon absorption. Phys. Rev. Lett. **65**, 96 (1990)

9. M. Sheik-Bahae, D.C. Hutchings, D.J. Hagan, E.W. van Stryland, Dispersion of bound electronic nonlinear refraction in solids. IEEE J. Quantum Electron. **27**, 1296 (1991)

10. A.M. Perelomov, V.S. Popov, M.V. Terent'ev, Ionization of atoms in an alternating electric field. Sov. Phys. JETP **23**, 924 (1966)

11. C. Brée, A. Demircan, G. Steinmeyer, Method for computing the nonlinear refractive index via Keldysh theory. IEEE J. Quantum Electron. **4**, 433 (2010)

12. H.A. Kramers, La diffusion de la lumiere par les atomes. Atti Cong. Intern. Fisica **2**, 545 (1927)

13. R. de L. Kronig. On the theory of the dispersion of X-rays. J. Opt. Soc. Am. **12**, 547 (1926)

14. D.C. Hutchings, M. Sheik-Bahae, D.J. Hagan, E.W. van Stryland, Kramers-Kronig relations in nonlinear optics. Opt. Quant. Electron. **24**, 1 (1992)

15. P. Lambropoulos, In *Topics on Multiphoton Processes in Atoms*, ed. by D. Bates, B. Bederson. Topics on Multiphoton Processes in Atoms, volume 12 of Advances in Atomic and Molecular Physics, (Academic Press, New York, 1976), pp. 87–164

16. P. Lambropoulos, X. Tang, Multiple excitation and ionization of atoms by strong laser fields. J. Opt. Soc. Am. B **4**, 821 (1987)

17. E. Varoucha, N.A. Papadogiannis, D. Charalambidis, A. Saenz, H. Schröder, B. Witzel, Quantitative laser mass spectroscopy of sputtered versus evaporated metal atoms. Phys. Rev. A **65**, 012901 (2001)

18. A. Apalategui, A. Saenz, Multiphoton ionization of the hydrogen molecule H_2. J. Phys. B: At. Mol. Opt. Phys. **35**, 1909 (2002)

19. L.V. Keldysh, Ionization in the field of a strong electromagnetic wave. Sov. Phys. JETP **20**, 1307 (1965)

20. F.H.M. Faisal, Collision of electrons with laser photons in a background potential. J. Phys. B: At. Mol. Phys. **6**, L312 (1973)

21. H.R. Reiss, Effect of an intense electromagnetic field on a weakly bound system. Phys. Rev. A **22**, 1786 (1980)

22. S.F.J. Larochelle, A. Talebpour, S.L. Chin, Coulomb effect in multiphoton ionization of rare-gas atoms. J. Phys. B **31**, 1215 (1998)

23. D.M. Volkov, Über eine Klasse von Lösungen der Diracschen Gleichung. Z. Phys. **94**, 250 (1935)

24. G.L. Yudin, M.Y. Ivanov, Nonadiabatic tunnel ionization: Looking inside a laser cycle. Phys. Rev. A **64**, 013409 (2001)

25. D. Bauer, D.B. Milosevic, W. Becker, Strong-field approximation for intense laser-atom processes: The choice of gauge. Phys. Rev. A **72**, 023415 (2005)

26. S.-I. Chu, W.P. Reinhardt, Intense field multiphoton ionization via complex dressed states: application to the H atom. Phys. Rev. Lett. **39**, 1195 (1977)

27. L. Pan, B. Sundaram, J. Lloyd Armstrong, Dressed-state perturbation theory for multiphoton ionization of atoms. J. Opt. Soc. Am. B **4**, 754 (1987)

28. W. Becker, A. Lohr, M. Kleber, Effects of rescattering on above-threshold ionization. J. Phys. B: At. Mol. Opt. Phys. **27**, L325 (1994)

29. E.U. Condon, G.H. Shortly, *Theory of Atomic Spectra*, (Cambridge University Press, Cambridge, 1935)
30. G. Simons, New model potential for pseudopotential calculations. J. Chem. Phys. **55**, 756 (1971)
31. L.P. Rapoport, B.A. Zon, N.L. Manakov, *Theory of Multiphoton Processes in Atoms (Section 2.4)*, (Atomizdat, Moscow, 1978)
32. V.S. Popov, Imaginary-time method in quantum mechanics and field theory. Phys. At. Nucl. **68**, 686 (2005)
33. R.P. Feynman, Space-time approach to non-relativistic quantum mechanics. Rev. Mod. Phys. **20**, 367 (1948)
34. P. Agostini, L.F. DiMauro, The physics of attosecond light pulses. Rep. Prog. Phys. **67**, 813 (2004)
35. F. Krausz, M. Ivanov, Attosecond physics. Rev. Mod. Phys. **81**, 163 (2009)
36. A.M. Peremolov, V.S. Popov, V.P. Kuznetsov, Allowance for the coulomb interaction in multiphoton ionization. Sov. Phys. JETP **27**, 451 (1968)
37. W. Becker, F. Grasbon, R. Kopold, D.B. Milosevic, G.G. Paulus, H. Walther, Above-threshold ionization: from classical features to quantum effects. Adv. At. Mol. Phys. **48**, 35 (2002)
38. M. Plummer, C.J. Noble, Calculations of resonance enhanced multiphoton ionization of argon in a KrF laser field. J. Phys. B: At. Mol. Opt. Phys. **33**, L807 (2000)
39. R. DeSalvo, A.A. Said, D.J. Hagan, E.W. van Stryland, M. Sheik-Bahae, Infrared to ultraviolet measurements of two-photon absorption and n2 in wide bandgap solids. IEEE J. Quantum Electron. **32**, 1324 (1996)
40. M. Sheik-Bahae, A.A. Said, T.H. Wei, D.J. Hagan, E.W. Stryland, Sensitive measurement of optical nonlinearities using a single beam. IEEE J. Quantum Electron. **26**, 760 (1990)
41. M. Nisoli, S.D. Silvestri, O. Svelto, R. Szipocs, K. Ferencz, C. Spielmann, S. Sartania, F. Krausz, Compression of high-energy laser pulses below 5fs. Opt. Lett. **22**, 522 (1997)
42. C.P. Hauri, W. Kornelis, F.W. Helbing, A. Heinrich, A. Couairon, A. Mysyrowicz, J. Biegert, U. Keller, Generation of intense, carrier-envelope phase-locked few-cycle laser pulses through filamentation. Appl. Phys. B **79**, 673 (2004)
43. G. Stibenz, N. Zhavoronkov, G. Steinmeyer, Self-compression of milijoule pulses to 7.8 fs duration in a white-light filament. Opt. Lett. **31**, 274 (2006)
44. H.J. Lehmeier, W. Leupacher, A. Penzkofer, Nonresonant third order hyperpolarizability of rare gases and N2 determined by third order harmonic generation. Opt. Commun. **56**, 67 (1985)
45. D.P. Shelton, J.E. Rice, Measurements and calculations of the hyperpolarizabilities of atoms and small molecules in the gas phase. Chem. Rev. **94**, 3 (1994)
46. J.E. Rice, Frequency-dependent hyperpolarizabilities for argon, krypton and neon: Comparison with experiment. J. Chem. Phys. **96**, 7580 (1992)
47. H.A. Lorentz, Über die Beziehung zwischen der Fortpflanzungsgeschwindigkeit des Lichtes und der Körperdichte. Ann. Phys. **9**, 641 (1880)
48. L. Lorenz, Über die Refractionsconstante. Ann. Phys. **11**, 70 (1880)
49. D.M. Bishop, J. Pipin, Improved dynamic hyperpolarizabilities and field-gradient polarizabilities for helium. J. Chem. Phys. **91**, 3549 (1989)
50. M.J. Shaw, C.J. Hooker, D.C. Wilson, Measurement of the nonlinear refractive index of air and other gases at 248 nm. Opt. Commun. **103**, 153 (1993)
51. J.-H. Klein-Wiele, T. Nagy, P. Simon, Hollow-fiber pulse compressor for KrF lasers. Appl. Phys. B **82**, 567 (2006)
52. J.T. Manassah, *Simple Models of Self-Phase and Induced-Phase Modulation*, The supercontinuum Laser Source, vol. 82 (Springer-Verlag, Berlin, 2006), p. 567
53. M.V. Ammosov, N.B. Delone, V.P. Krainov, Tunnel ionization of complex atoms and of atomic ions in an alternating electromagnetic field. Sov. Phys. JETP **64**, 1191 (1986)
54. V.S. Popov, Tunnel and multiphoton ionization of atoms and ions in a strong laser field. Phys. Usp. **47**, 855 (2004)
55. T. Brabec, F. Krausz, Intense few-cycle laser fields: frontiers of nonlinear optics. Rev. Mod. Phys. **72**, 545 (2000)

56. R.W. Boyd, *Nonlinear Optics*, (Academic Press, Orlando, 2008)
57. L. Berge, S. Skupin, R. Nuter, J. Kasparian, J.P. Wolf, Ultrashort filaments of light in weakly ionized, optically transparent media. Rep. Prog. Phys. **70**, 1633 (2007)
58. F.W.J. Olver, D.W. Lozier, R.F. Boisvert, C.W. Clark (eds.), *NIST Handbook of Mathematical Functions* (Cambridge University Press, Cambridge, 2010)
59. D.M. Bishop, Dispersion formulas for certain nonlinear optical processes. Phys. Rev. Lett. **61**, 322 (1988)
60. J.W.N. Thomas Lundeen, S-Y. Hou, Nonresonant third order susceptibilities for various gases. J. Chem. Phys. **79**, 6301 (1983)
61. P. Fuentealba, O. Reyes, Polarizabilities and hyperpolarizabilities of the alkali metal atoms. J. Phys. B: At. Mol. Opt. Phys. **26**, 2245 (1993)
62. T. Kobayashi, K. Sasagane, K. Yamaguchi, Frequency-dependent second hyperpolarizabilities in the time-dependent restricted open-shell Hartree-Fock theory: application to the Li, Na, K, and N atoms. J. Chem. Phys. **112**, 7903 (2000)
63. J. Stiehler, J. Hinze, Calculation of static polarizabilities and hyperpolarizabilities for the atoms He through Kr with a numerical RHF method. J. Phys. B: At. Mol. Opt. Phys. **28**, 4055 (1995)
64. C. Lupinetti, A.J. Thakkar, Polarizabilities and hyperpolarizabilities for the atoms Al, Si, P, S, Cl, and Ar: coupled cluster calculations. J. Chem. Phys. **122**, 044301 (2005)
65. Y. Malykhanov, I. Eremkin, S. Begeeva, Calculation of the dipole hyperpolarizability of atoms with closed and open shells by the Hartree-Fock-Roothaan method. J. Appl. Spectrosc. **75**, 1 (2008). ISSN 0021-9037.10.1007/s10812-008-9020-y
66. M.R. Junnarkar, N. Uesugi, Near two-photon resonance short pulse compression in atomic noble gases. Opt. Commun. **175**, 447 (2000)
67. M.J. DeWitt, E. Wells, R.R. Jones, Ratiometric comparison of intense field ionization of atoms and diatomic molecules. Phys. Rev. Lett. **87**, 153001 (2001)
68. A. Talebpour, J. Yang, S.L. Chin, Semi-empirical model for the rate of tunnel ionization of N2 and O2 molecule in an intense Ti:sapphire laser pulse. Opt. Commun. **163**, 29 (1999)
69. R.W. Hellwarth, D.M. Pennington, M.A. Henesian, Indices governing optical self-focusing and self-induced changes in the state of polarization in N_2, O_2, H_2, and Ar gases. Phys. Rev. A **41**, 2766 (1990)
70. E.T.J. Nibbering, G. Grillon, M.A. Franco, B.S. Prade, A. Mysyrowicz, Determination of the inertial contribution to the nonlinear refractive index of air, N2, and O2 by use of unfocused high-intensity femtosecond laser pulses. J. Opt. Soc. Am. B **14**, 650 (1997)
71. W. Liu, High-order nonlinear susceptibilities of helium. Phys. Rev. A **56**, 4938 (1997)
72. M. Nurhuda, A. Suda, K. Midorikawa, Ionization-induced high-order nonlinear susceptibility. Phys. Rev. A **66**, 041802(R) (2002)
73. M. Nurhuda, A. Suda, K. Midorikawa, Saturation of dynamic nonlinear susceptibility of noble gas atoms in intense laser field. RIKEN Rev. **48**, 40 (2002)
74. C. Brée, A. Demircan, G. Steinmeyer, Kramers-Kronig relations and high-order nonlinear susceptibilities. Phys. Rev. A **85**, 033806 (2012). doi:10.1103/PhysRevA.85.033806
75. V. Loriot, P. Bãªjot, W. Ettoumi, Y. Petit, J. Kasparian, S. Henin, E. Hertz, B. Lavorel, O. Faucher, J. Wolf, On negative higher-order Kerr effect and filamentation. Laser Phys. **21**, 1319 (2011). ISSN 1054–660X.10.1134/S1054660X11130196

Chapter 5
Conclusions

In the present thesis, femtosecond filamentation was investigated. While the first part of this work explores the self-compression both, theoretically and experimentally, and reveals the physical mechanisms behind this remarkable phenomenon, the second part affects the foundations of femtosecond filamentation and, moreover, those of nonlinear optics at extreme intensities. A totally new approach for a theoretical prediction of the magnitude of the higher-order nonlinear susceptibilties is presented, which is in excellent agreement with recent experimental results [1].

In Chap. 3, filamentary self-compression is traced back to a self-pinching mechanism which can be regarded as analogous to the z-pinch of magnetohydrodynamics [2, 3]. It was shown that the interplay of purely spatial effects, i.e., Kerr self-focusing and plasma defocusing can lead to a considerable dynamics of the temporal pulse profile, which is related to the noninstantaneous nature of the plasma nonlinearity. This temporal dynamics involves temporal splittings of the pulse, as substantiated by a simple analytical model. Under suitable input pulse conditions, the plasma-induced pulse splitting may introduce a split-isolation cycle that yields a few-cycle self-compressed pulse. Moreover, the latter results revealed that the characteristic longitudinal structure of a filament, with a strongly ionized zone followed by a nearly plasmaless, subdiffractive channel [4] already appears in a purely spatial model, neglecting any temporal effects like dispersion and self-steepening. This further corrobarates that the prevalent mechanisms behind filamentary self-compression are of purely spatial nature.

In the postionization zone [5], it was shown that refocusing events, partially arrested by GVD, can lead to a second split-isolation cycle which cascades the self-compression mechanism and strongly increases the compression ratio. As both experimental and numerical studies revealed [6], refocusing events can give rise to self-compressed few-cycle pulses with characteristic temporal and spectral signatures evident in a spectrogram representation. A further intriguing feature of femtosecond filaments is their ability to restore both, their spatial and temporal profile. The former was evidenced in [7], where it was shown that a filament can self-restore its transverse spatial profile after hitting an obscurant with a diameter of up to 2/3

C. Brée, *Nonlinear Optics in the Filamentation Regime*, Springer Theses, 111
DOI: 10.1007/978-3-642-30930-4_5, © Springer-Verlag Berlin Heidelberg 2012

of the filament core, and it is suggested that "The replenishment of the pulse mainly proceeds from the nonlinear attractor responsible for the formation of a spatial soliton modeling the filament core." [7], a proposal which is substantiated by showing that a time-averaged, approximate analogue of the underlying dynamical equations indeed admits stable soliton solutions. Temporal self-restoration has been theoretically predicted in [8, 9] by means of the impact of a thin window of fused silica on the filamentary self-compressed pulse. In a typical experimental setup, the self-compressed optical pulses leave the gas cell by traversing such a silica window. The results presented in Sect. 3.4. successfully evidenced these theoretical predictions, and revealed a strong impact of the longitudinal position of the exit window. In particular, the measurements and numerical simulations also revealed that temporal self-restoration may become ineffective for an unsuitably positioned gas cell. This is further corroborated by measurements with a window-less gas cell. These findings might be useful for increasing the efficiency of filamentary self-compression in future experiments.

Finally, a possible shift of paradigm in the field of femtosecond filamentation and maybe nonlinear optics as a whole, has been indicated in Chap. 4. A theoretical model explaining the saturation and inversion behaviour of the IDRI was presented, based on Kramers-Kronig relations for the nonlinear optical susceptibilities, cf. Eq. (4.58). This model was originally developed to calculate the second-order Kerr coefficient n_2 of semiconductors [10, 11] and later applied to noble gases [B5]. In Sect. 4.3, this model is extended to yield more accurate predictions on the dispersion of n_2 with wavelength. The obtained results show excellent agreement with independent theoretical and experimental data [B9]. Having benchmarked the Kramers-Kronig approach on the basis of the second-order nonlinear refractive index, the method was generalized to calculate higher-order Kerr terms, yielding favorable agreement with experimental measurements of Loriot et al. [1, 12].

Nonlinear refraction may be understood to arise from virtual multiphoton transitions from the ground state to an excited bound or continuum state and back to the ground state. This sequence of transitions produces a phase-shift of the participating photons which, on a macroscopic level, is responsible for the observed self-phase modulation due to second- or higher order Kerr nonlinearities. As the employed ionization model [13] disregards the existence of internal atomic resonances, the Kramers-Kronig approach of Chap. 4 captures only the contribution of transitions between the ground and the continuum states to nonlinear refraction. An alternative approach to analyze the contribution of bound-continuum transitions to the IDRI was presented in Ref. [14]. In the latter work, the IDRI of a simplified model system sometimes referred to as 'delta-Hydrogen' was calculated. This system uses a delta function to approximate the atomic potential. The latter potential only admits a single bound-state, such that in this model, similar to the approach employed here, only bound-continuum transitions contribute to the nonlinear susceptibility. In fact, within the 'delta-Hydrogen' model, the IDRI exhibits qualitatively the same saturation and inversion behavior as that observed in [1, 12] and the present thesis. Nevertheless, the authors of Ref. [14] argue, in support of the classical model of filamentation, that in real atomic systems, virtual transitions between bound states provide the dominant contribution to the IDRI, therefore masking the saturation behavior due to the continuum transitions. In contrast, the remarkable agreement obtained in Sect. 4.3. of the second-order nonlinear refractive index n_{2k} with independent experimental and theoretical data, which is absent in Ref. [14], indeed suggests that transitions between the ground state and the continuum states provide a considerable contribution to nonlinear refraction. Nevertheless, a future refinement of the model may involve to augment the multiphoton cross-sections β_K (cf. Eq. (4.55)) with terms which take into account participating excited bound states, in order to account for multiphoton transitions between bound states or resonance enhanced MPI.

An experiment to provide an independent test of the predictions of Loriot et al. was proposed in Ref. [15]. This involves measuring the yield of third compared to fifth-harmonic generation in the gas under consideration. Indeed, according to the analysis of Ref. [16], the experimental results of [17] for argon are in support of the higher-order Kerr model.

These results strongly suggest to include a Kerr-based saturation mechanism into future models of filament formation. Modeling of white-light propagation, however, may turn out to be difficult because of the strong dispersion of the higher-order coefficients, and methods for efficient modeling of dispersive nonlinearities may have to be found.

References

1. V. Loriot, E. Hertz, O. Faucher, B. Lavorel, Measurement of high order Kerr refractive index of major air components. Opt. Express **17**, 13429 (2009)
2. V.F. D'Yachenko, V.S. Imshenik, Magnetohydrodynamic theory of the pinch effect in a dense high-temperature plasma (dense plasma focus). Rev. Plasma Phys. **5**, 447 (1970)
3. J.B. Taylor, Relaxation of toroidal plasma and generation of reverse magnetic fields. Phys. Rev. Lett. **33**, 1139 (1974)
4. S.L. Chin, Y. Chen, O. Kosareva, V.P. Kandidov, F. Théberge, What is a filament? Laser Phys. **18**, 962 (2008)
5. S. Champeaux, L. Bergé, Postionization regimes of femtosecond laser pulses self-channeling in air. Phys. Rev. E **71**, 046604 (2005)
6. C. Brée, J. Bethge, S. Skupin, L. Bergé, A. Demircan, G. Steinmeyer, Cascaded self-compression of femtosecond pulses in filaments. New J. Phys. **12**, 093046 (2010)
7. S. Skupin, L. Bergé, U. Peschel, F. Lederer, Interaction of femtosecond light filaments with obscurants in aerosols. Phys. Rev. Lett. **93**, 023901 (2004)
8. L. Berge, S. Skupin, G. Steinmeyer, Temporal self-restoration of compressed optical filaments. Phys. Rev. Lett. **101**, 213901 (2008)
9. L. Bergé, S. Skupin, G. Steinmeyer, Self-recompression of laser filaments exiting a gas cell. Phys. Rev. A **79**, 033838 (2009)
10. M. Sheik-Bahae, D.J. Hagan, E.W. van Stryland, Dispersion and band-gap scaling of the electronic Kerr effect in solids associated with two-photon absorption. Phys. Rev. Lett. **65**, 96 (1990)
11. M. Sheik-Bahae, D.C. Hutchings, D.J. Hagan, E.W. van Stryland, Dispersion of bound electronic nonlinear refraction in solids. IEEE J. Quantum Electron. **27**, 1296 (1991)
12. V. Loriot, E. Hertz, O. Faucher, B. Lavorel, Measurement of high order Kerr refractive index of major air components: erratum. Opt. Express **18**, 3011 (2010)

13. V.S. Popov, Tunnel and multiphoton ionization of atoms and ions in a strong laser field. Phys. Usp. **47**, 855 (2004)
14. A. Teleki, E.M. Wright, M. Kolesik, Microscopic model for the higher-order nonlinearity in optical filaments. Phys. Rev. A **82**, 065801 (2010)
15. M. Kolesik, E.M. Wright, J.V. Moloney, Femtosecond filamentation and higher-order nonlinearities. Opt. Lett. **35**, 2550 (2010)
16. P. Béjot, E. Hertz, B. Lavorel, J. Kasparian, J.-P. Wolf, O. Faucher, From higher-order Kerr nonlinearities to quantitative modeling of third and fifth harmonic generation in argon. Opt. Lett. **36**, 828 (2011)
17. K. Kosma, S.A. Trushin, W.E. Schmid, W. Fuß, Vacuum ultraviolet pulses of 11 fs from fifth-harmonic generation of a Ti:sapphire laser. Opt. Lett. **33**, 723 (2008)

Appendix A
The Nonlinear Schrödinger Equation

The NLSE (2.73) describes the self-focusing of optical beams in a nonlinear Kerr medium, a phenomenon which is embedded into the more general context of wave collapse and self-focusing [1]. Therefore, both from a physical and functional analytic point of view, it is worthwhile to study a generalized NLSE in one longitudinal dimension η (the propagation direction) and D transverse dimensions parametrized by $(\xi_1, \ldots, \xi_D)$ according to

$$i\partial_\eta \psi + \sum_{j=1}^{D} \frac{\partial^2}{\partial \xi_j^2} \psi(\eta; \vec{\xi}) + |\psi|^{2\sigma}\psi = 0 \tag{A.1}$$

As the existence of stable, localized structures of wave energy is of considerable interest from the viewpoint of technological applications, a good deal of the mathematical theory on the NLSE is devoted to the existence of standing wave solutions of the form

$$\Phi(\eta, \vec{\xi}) = R(\vec{\xi})e^{i\lambda\eta}. \tag{A.2}$$

Given that the standing wave solution is stable under infinitesimal perturbations, it is also referred to as soliton solution of the NLSE. Inserting Eq. (A.2) into the NLSE (1), one obtains the PDE

$$\sum_{j=1}^{D} \frac{\partial^2}{\partial \xi_j^2} R - \lambda R + |R|^{2\sigma}R = 0. \tag{A.3}$$

For $\sigma = D = 1$, the latter equation can be solved analytically, yielding

$$R_\lambda(\xi) = \frac{\sqrt{2\lambda}}{\cosh(\sqrt{\lambda}\xi)}. \tag{A.4}$$

This is the well known fundamental soliton of the NLSE, which has found widespread applications in nonlinear fiber optics [2, 3]. A stability criterion for stationary states of the NLSE under infinitesimal perturbations has been derived in

C. Brée, *Nonlinear Optics in the Filamentation Regime*, Springer Theses,
DOI: 10.1007/978-3-642-30930-4, © Springer-Verlag Berlin Heidelberg 2012

Refs. [4–7]. Introducing the soliton mass

$$N(\lambda) = \int d^D \xi |R_\lambda(\vec{\xi})|^2, \tag{A.5}$$

stable ground states exist whenever

$$\frac{d}{d\lambda} N(\lambda) > 0. \tag{A.6}$$

This condition for the existence of stable solitons is also known as Vakhitov-Kolokolov criterion. It can be shown that it is equivalent to the condition $\sigma D < 2$, which specifies the so-called subcritical case, while $D\sigma = 2$ and $D\sigma > 2$ are referred to as critical and supercritical cases. The critical case is of special relevance for the physics of intense optical beams. Letting $D = 2$ and $\sigma = 1$, it can be shown that the ground-state solution $R_{0,\lambda}$ to Eq. (A.3) is a continuous, positive function with no nodes. Furthermore, it is radially symmetric, $R_{0,\lambda} = R_{0,\lambda}(\rho)$, where $\rho = \sqrt{\xi_1^2 + \xi_2^2}$. The class of solutions $R_{0,\lambda}$ to Eq. (A.3) for arbitrary λ can be generated from the solution $R_{0,1}$ for $\lambda = 1$ by means of the scale transformation

$$R_{0,\lambda}(\rho) = \sqrt{\lambda} R_{0,1}(\sqrt{\lambda}\rho). \tag{A.7}$$

From this scaling property it is straightforward to conclude that in the critical case, the mass of the ground-state $N(\lambda) = 2\pi \int d\rho \rho R_{0,\lambda}^2(\rho)$ is independent of λ, i.e., $N(\lambda) = N(1)$. Therefore, it follows from the Vakhitov-Kolokolov criterion (6) that the ground-state solution R_0 is unstable, and it can be shown that for any $\epsilon > 0$, there exists an initial condition $\phi(0, \vec{\xi})$ of the NLSE close to R_0 with $\|\phi - R_0\| < \epsilon$ such that the amplitude of $\phi(\eta, \vec{\xi})$ blows up at finite distance η_c, i.e., $\lim_{\eta \to \eta_c} \phi(\eta, \vec{\xi}) = \infty$. Moreover, the groundstate mass $N_0 = \|R_{0,1}\|^2$ provides a lower bound for the threshold mass required to observe finite-distance blowup: any initial datum ϕ with $\|\phi\|^2 < N_0$ evolves into a globally defined solution to the NLSE and does not blow up. Therefore, a necessary condition for blow up is provided by

$$\|\phi^2\| > N_0 \tag{A.8}$$

The ground state $R_{0,1}$ is also known as the Townes mode [8]. It was obtained by numerically solving Eq. (A.3) for $D = 2$ and $\sigma = 1$, leading to $N_0 \approx 11.69$. In the latter reference, the Townes mode is discussed in the context of optical self-focusing and self-trapping of optical beams in a self-generated waveguide. Indeed, the paraxial wave Eq. (2.73) in a self-focusing nonlinear Kerr medium with $n(I) = n_0 + n_2 I$ is recovered after the substitutions $4z_0\eta \to z$, $w_0(\xi_1, \xi_2) \to (x, y)$ and $\sqrt{c_2}\psi \to \mathcal{E}$, where $c_2 = \lambda^2/(8\pi^2 n_0 n_2 w_0^2)$. Adopting the above results to optical self-focusing, it is found that in analogy to Eq. (A.8), self-focusing and blow up at the critical distance provided by the Marburger formula (2.76) is possible only under the necessary condition that the optical power P of the input beam exceeds the critical power $P_{cr} = N_0\lambda^2/(8\pi^2 n_0 n_2)$, cf. Eq. (2.75).

An important analytical tool for the mathematical analysis of self-focusing and blow up is provided by the so-called virial identity. Introducing the variance V and Hamiltonian H of the wavefield ψ according to

$$V = \int |\vec{\xi}|^2 |\psi|^2 d^2\xi \tag{A.9}$$

$$H = \int d^2\xi \left(|\vec{\nabla}\psi|^2 - \frac{1}{2}|\psi|^4 \right), \tag{A.10}$$

it is found that any solution to the NLSE 1 satisfies

$$\partial_\eta^2 V = 8H. \tag{A.11}$$

Here, only the case $(D = 2, \sigma = 1)$ is considered. As stationary solutions to the NLSE have constant variance, it immediately follows from the virial identity that $H = 0$. Therefore, in critical dimension $D = 2$, $\sigma = 1$, the Townes mode is a zero-energy solution of the NLSE. Using the virial identity, an estimate for the critical power can be obtained. As a trial function, a Gaussian beam of waist $\Pi(\eta)$ and optical power $\tilde{P}$ in dimensionless units, is chosen according to

$$\psi(\eta, \vec{\xi}) = \sqrt{\frac{\tilde{P}}{\pi\varrho^2(\eta)}} \exp\left(-\frac{\rho^2}{2\Pi^2(\eta)} + i\frac{1}{4}\frac{\Pi_\eta(\eta)\rho^2}{\Pi(\eta)} \right) \tag{A.12}$$

Inserting this ansatz into the virial identity 11 yields an ordinary differential equation that governs the evolution of the beam waist Π,

$$\frac{1}{4}\Pi^3(\eta)\frac{d^2}{d\eta^2}\Pi(\eta) = 1 - \frac{\tilde{P}}{\tilde{P}_{cr}} \tag{A.13}$$

For linear diffraction, the right hand side of this equation is equal to unity. However, for optical beams with a power exceeding the critical power $\tilde{P}_{cr}$, the right-hand side reverses the sign compared to linear theory, and the beam waist eventually contracts to zero, yielding an intensity blow-up at some finite distance η_*. The evolution equation for the beam waist can be solved analytically, and reverting to physical units, the solution is given by Eq. (3.11) of Sect. 3.11, with a critical power approximated by means of the variational approach according to

$$P_{cr} = \frac{\lambda^2}{2\pi n_0 n_2}, \tag{A.14}$$

References

[1] C. Sulem, P.-L. Sulem (1999) *The Nonlinear Schrödinger Equation: Self-Focusing and Wave Collapse. Applied Mathematical Sciences*, Springer

[2] G.P. Agrawal (2001) *Nonlinear Fiber Optics*. Academic Press, 3rd edn

[3] F. Mitschke (2009) *Fiber Optics: Physics and Technology*. Springer, 1st edn

[4] M.I. Weinstein (1986) Lyapunov stability of ground states of nonlinear dispersive evolution equations. Commun. Pure and Applied Math., **39**:51

[5] M. Grillakis, J. Shatah, W.A. Strauss (1987) Stability theory of solitary waves in the presence of symmetry, I. J. Funct. Anal., **74**:160

[6] N.G. Vakhitov, A.A. Kolokolov (1973) Stationary solutions of the wave equation in a medium with nonlinearity saturation. Izv. Vuz. Radiofiz., **16**:1020

[7] A.A Kolokolov (1976) Stability of stationary solutions of nonlinear wave equations. Izv. Vuz. Radiofiz., **17**:1016

[8] R.Y. Chiao, E. Garmire, C.H. Townes, (1964) Self-Trapping of Optical Beams. Phys. Rev. Lett., **13**:479

Appendix B
Numerical Method

The FORTRAN90 code implementing a numerical integration scheme for the set of Eqs. (2.55), (2.56) was kindly provided by Stefan Skupin (MPIPKS) and Luc Bergé (CEA-DAM). The code uses a split-step pseudospectral method [1] to integrate the envelope Eq. (2.55), which can be represented as

$$-i\partial_z\hat{\mathcal{E}} = \mathcal{L}(\omega)\mathcal{E} + \mathcal{N}(\hat{\mathcal{E}}, \omega)$$
(B.15)

where the operator $\mathcal{L}$ given by

$$\mathcal{L}(\omega) = \mathcal{L}_r(\omega) + \mathcal{D}(\omega)$$
(B.16)

subsumes the linear optical effects. It can further be decomposed into a contribution $\mathcal{D}(\omega)$ Eq. (2.59) modeling temporal dispersion and a radial part $\mathcal{L}_r$,

$$\mathcal{L}_r(\omega) = \frac{1}{2k_0}\,\widehat{T}^{-1}(\omega)\frac{1}{r}\partial_r r \partial_r$$
(B.17)

which models linear diffraction within the paraxial approximation, including space-time focusing effects. Here, $\widehat{T} = 1 + \omega/\omega_0$ is the frequency domain representation of the operator T, Eq. (2.57). The nonlinear propagation effects are contained in $\mathcal{N}(\hat{\mathcal{E}}, \omega)$, which is given by

$$\mathcal{N}(\hat{\mathcal{E}}, \omega) = i\frac{\omega_0}{c}n_2\hat{T}(\omega)\mathcal{F}\left[|\mathcal{E}|^2\mathcal{E}\right](\omega) - i\frac{k_0}{2\rho_c}\hat{T}^{-1}(\omega)\mathcal{F}[\rho(\mathcal{E})\mathcal{E}](\omega)$$
$$- \frac{\sigma}{2}\mathcal{F}[\rho\mathcal{E}](\omega) - \mathcal{F}\left[\frac{U_iW(I)(\rho_{nt} - \rho)}{2I}\mathcal{E}\right](\omega)$$
(B.18)

The applied pseudospectral method numerically integrates Eq. (2.55), obtaining the envelope $\hat{\mathcal{E}}(r, \omega, z_0 + \Delta z)$ from given initial datum $\hat{\mathcal{E}}(r, z_0, \omega)$, by first advancing the initial datum along a distance $\Delta z/2$ according to the equation obtained by setting $\mathcal{N} = 0$ in Eq. (B.15). Formally, the solution to the corresponding linear evolution equation may be written

C. Brée, *Nonlinear Optics in the Filamentation Regime*, Springer Theses,
DOI: 10.1007/978-3-642-30930-4, © Springer-Verlag Berlin Heidelberg 2012

$$\hat{\mathcal{E}}(r, z + \Delta z/2, \omega) = e^{i\frac{\Delta z}{2}\mathcal{L}_r(\omega)} e^{i\frac{\Delta z}{2}\mathcal{D}(\omega)} \hat{\mathcal{E}}(r, z, \omega) \tag{B.19}$$

as the operators $\mathcal{L}_r(\omega)$ and $\mathcal{D}(\omega)$ commute with each other. While a solution to the purely dispersive part of the evolution equation is readily obtained by multiplication of the initial datum with the phase factor $\exp(i\Delta z\mathcal{D}(\omega)/2)$, the action of the radial operator $\exp(i\Delta z\mathcal{L}_r(\omega)/2$ is approximated numerically by using a finite-difference, implicit Crank-Nicholson scheme [2] with transparent boundary conditions in r [3].

The linearly evolved initial datum is then further propagated along a distance Δz according to the nonlinear equation obtained by letting $\mathcal{L} = 0$ in Eq. (B.15). The nonlinear integration step is performed using a Runge-Kutta method. However, in the frequency domain, terms nonlinear in the envelope correspond to multifold convolutions, which can only be performed at large computational cost. Therefore, prior to performing the nonlinear step, the fast fourier transform method (FFT) is applied on $\hat{\mathcal{E}}(r, \omega, z)$ to obtain the time-domain representation of the envelope $\mathcal{E}(r, z, t)$, which enables a straightforward computation of the nonlinear terms. The aliasing error introduced by this procedure is controlled by applying a low-pass filter on $\hat{\mathcal{E}}$ in the frequency domain. Furthermore, usage of the FFT implies imposing periodic boundary conditions on the envelope both in the time and frequency domain.

A repeated linear propagation step along a distance $\Delta z/2$ completes the integration scheme, yielding the electric field envelope $\mathcal{E}(r, z + \Delta z, t)$ from an initial envelope $\mathcal{E}(r, z, t)$, subject to the full propagation model Eq. (B.15).

The numerical computations are parallelized using the Message Passing Interface (MPI) libraries and executed on the WIAS blade cluster `euler` (Hewlett-Packard CP3000BL). The cluster consists of 32 blades of the type HP BL460c and 16 blades of type HP BL2x220c. Each blade is equipped with two INTEL Xeon5430/2666 Quad Core processors and provides 16 GB of RAM. In order to implement an efficient parallelization of the employed pseudospectral method, the two-dimensional numerical grid has to be distributed among the compute nodes. It is assumed that the discretized r and t coordinates label columns and rows of the grid, respectively. Then, the conduction of the FFT requires each compute node to store entire columns, while for the application of the Crank-Nicholson method, each node has to store entire rows. Therefore, the distribution of the grid among the compute nodes has to be transposed repeatedly. An efficient method to perform the necessary data exchange between the nodes is described in [4].

References

[1] T.R. Taha, M.I. Ablowitz, (1984) Analytical and numerical aspects of certain nonlinear evolution equations. II. Numerical, nonlinear Schrödinger equation. J. Comput. Phys., **55**:203

[2] J. Crank, P. Nicolson, (1996) A practical method for numerical evaluation of solutions of partial differential equations of the heat-conduction type. Adv. Comput. Mathe., **6**, 207 ISSN 1019-7168. 10.1007/BF02127704

[3] G.R. Hadley, (1991) Transparent boundary condition for beam propagation. Opt. Lett., **16**:624

[4] S. Skupin (2005) Nonlinear dynamics of trapped beams. Ph.D. thesis, Friedrich-Schiller-Universität Jena

Appendix C
Characterization of Ultrashort
Few-Cycle Pulses

Spectral phase interferometry for direct electric field reconstruction (SPIDER) [1] is an interferometric method to characterize the spectral phase of ultrashort few-cycle optical pulses. The spectral phase $\phi(\omega)$ is defined as the phase of the electric field $E(t)$ in the frequency domain, i.e., as the argument of the Fourier transform $\hat{E}(\omega)$ of $E(t)$,

$$\hat{E}(\omega) = |\hat{E}(\omega)|e^{i\phi(\omega)}. \tag{C.20}$$

In principle, the spectral phase of an optical pulse can be reconstructed by means of the Takeda algorithm, from the interference signal generated by two spectrally sheared replica of the pulse. In practice, the spectrally sheared replica are generated by frequency up-conversion in a $\chi^{(2)}$ medium, see Fig. C.1. To this purpose, a thin glass plate is used as an etalon. The front and back reflex off the etalon provides two copies of the pulse with mutual temporal delay $\Delta\tau$. In addition, the fraction of the pulse transmitted through the glass plate is sent through a dispersive medium, e.g., BK7 glass. The acquired frequency chirp stretches the pulse temporally and maps its frequency content into the time domain. Subsequently, the chirped pulse and the two copropagating copies of the pulse are subject to sum frequency generation in a $\chi^{(2)}$ medium. In the experiments of Sect. 3.4., a BBO crystal under Type-II phase matching conditions was used. As the chirped pulse is sufficiently long, the SFG process may be described by resulting from the interaction of a short pulse with a monochromatic wave, giving rise to a frequency up-conversion of the short pulses. However, due to the frequency chirp of the stretched pulse and the temporal delay between the short pulses, each of them is upconverted with a slightly different respective frequency. The SPIDER signal thus consists of two spectrally sheared pulses with a frequency offset $\Delta\omega$ that is determined by the delay $\Delta\tau$ and the GDD of the chirped pulse. In the frequency domain, it is given by

$$S(\omega) = \left|\hat{E}(\omega - \omega_0)e^{i(\omega-\omega_0)\Delta\tau} + \hat{E}(\omega - \omega_0 - \Delta\omega)\right|^2. \tag{C.21}$$

C. Brée, *Nonlinear Optics in the Filamentation Regime*, Springer Theses,
DOI: 10.1007/978-3-642-30930-4, © Springer-Verlag Berlin Heidelberg 2012

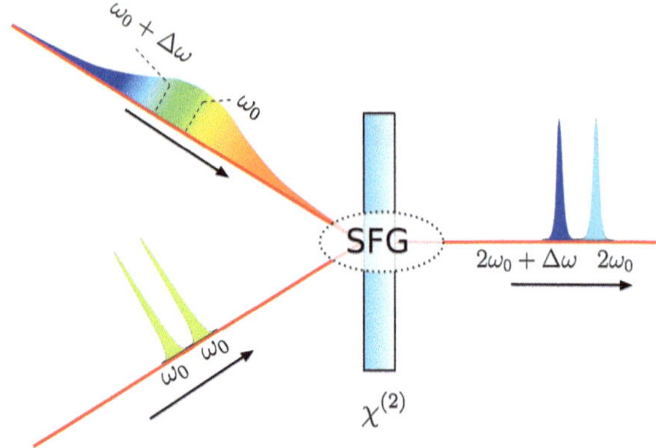

Fig. C.1 Schematic representation of a experimental SPIDER setup. Two mutually delayed copies of the pulse to be characterized are subject to sum frequency generation (SFG) in a $\chi^{(2)}$-crystal with a third, chirped copy of the pulse in a Type-II phase matching geometry

With the polar decomposition Eq. (C.20), this may be evaluated to give

$$
\begin{aligned}
S(\omega) =& \left|\hat{E}(\omega')\right|^2 + \left|\hat{E}(\omega' - \Delta\omega)\right|^2 + 2\left|\hat{E}(\omega')\hat{E}(\omega' - \Delta\omega)\right| \times \\
& \times \cos\left(\phi(\omega') - \phi(\omega' - \Delta\omega) + \phi_{\text{ref}}(\omega')\right),
\end{aligned}
\tag{C.22}
$$

with a linear *reference phase* $\phi_{ref} = \omega'\Delta\tau$. Experimentally, the SPIDER signal $S(\omega)$ is obtained by analyzing the copropagating spectrally sheared pulses with a spectrograph. This gives rise to a characteristic fringe pattern, as displayed in Fig. 3.18a. A close inspection of Eq. (C.22) reveals that (for nontrivial spectral phase) the fringe spacing is not constant along ω. In fact, the phase of the cosine in Eq. (C.22) consists of a linear contribution $\omega'\Delta\tau$ plus a phase modulation due to the presence of the finite difference

$$
\theta(\omega') = \phi(\omega') - \phi(\omega' - \Delta\omega).
\tag{C.23}
$$

By analyzing the modulation of the fringe spacing and after subtracting the linear reference phase $\phi_{ref} = \omega'\Delta\tau$, the group delay of the pulse is obtained, where it is assumed that

$$
\theta(\omega') \approx \Delta\omega \frac{\partial\phi(\omega')}{\partial\omega'}.
\tag{C.24}
$$

The necessary phase demodulation can, e.g., be obtained by using the Takeda algorithm [2] or wavelet based demodulation strategies [3, 4].

The reference phase ϕ_{ref} can, in principle, be obtained without further measurement by calculating the delay $\Delta\tau$ from the thickness of the glass etalon. In practice, however, the spectral phases of copropagating pulses do not coincide exactly, as the pulse resulting from the back reflex of the etalon is subject to additional dispersive shaping while traversing the etalon. Therefore, under experimental conditions, the reference phase is quasi-linear, $\phi_{\text{ref}}(\omega') = \omega'\Delta\tau + \xi(\omega')$, with a small deviation $\xi(\omega')$ which accounts for the dispersive effects within the SPIDER apparatus. To minimize this effect, the reference phase can be measured by screening off the chirped pulse and changing the orientation of the nonlinear crystal to achieve Type I phase matching. This enables the copropagating pulses to generate a second harmonic signal within the BBO crystal. The SHG signal is subsequently recorded by a spectrograph and the resulting fringe pattern is demodulated, in analogy to the demodulation of the SPIDER signal. This procedure yields the required reference phase ϕ_{ref}, which is subtracted from the demodulated SPIDER phase to yield the group delay of the input pulse. The spectral phase can finally be obtained by integrating Eq. (C.24),

$$\phi(\omega') = \frac{1}{\Delta\omega} \int d\omega' \theta(\omega') + C_0 \tag{C.25}$$

As the SPIDER method does not provide experimental means to determine the integration constant C_0, the electric field is determined up to a constant phase shift, cf. (C.20). Therefore, SPIDER is a technique to determine the *envelope* of an ultrashort pulse and is insensitive to an offset between the carrier signal and the envelope.

While SPIDER is an interferometric method, the XFROG method is non-interfero-metric. It relies on the generation of a cross-correlation signal generated by focusing the test pulse $E(t)$ to be characterized and a well-characterized reference pulse $E_{\mathrm{ref}}(t)$ into a $\chi^{(2)}$ medium, e.g., a BBO crystal. Furthermore, a delay stage is used to introduce a variable temporal delay τ of the reference pulse w.r.t. the test pulse. The electric field of the cross-correlation signal detected behind the BBO crystal is then given by

$$E_X(t, \tau) = E(t)E_{\mathrm{ref}}(t - \tau). \tag{C.26}$$

When the XFROG signal is analyzed with a spectrograph, the spectral intensity of the XFROG signal provides the XFROG trace,

$$I_X(\omega, \tau) = \left| \int_{-\infty}^{\infty} dt E_X(t, \tau) e^{i\omega\tau} \right|^2. \tag{C.27}$$

References

[1] C. Iaconis, I.A. Walmsley (1998) Spectral phase interferometry for direct electric-field reconstruction of ultrashort optical pulses. Opt. Lett., **23**:792
[2] M. Takeda, H. Ina, S. Kobayashi (1982) Fourier-transform method of fringe-pattern analysis for computer-based topography and interferometry. J. Opt. Soc. Am., **72**:156
[3] J. Bethge, G. Steinmeyer (2008) Numerical fringe pattern demodulation strategies in interferometry. Rev. Sci. Instrum., **79**:073102
[4] J. Bethge, C. Grebing, G. Steinmeyer, (2007) A fast Gabor wavelet transform for high-precision phase retrieval in spectralinterferometry. Opt. Express, **15**:14313